Writing Scientific Papers and Reports

Writing Scientific Papers and Reports

Eighth Edition

W. Paul Jones
Iowa State University
Ames

Revised by
Michael L. Keene
Texas A & M University
College Station

wcb

Wm. C. Brown Company Publishers

Dubuque, Iowa

Cover photo by Ken Smith

Library of Congress Catalog Card Number: 81–65432

ISBN 0–697–03773–8

Printed in the United States of America
10 9 8 7 6 5 4 3

Contents

Documentation, Mechanics, Diction

Preface

For more than thirty years W. Paul Jones's *Writing Scientific Papers and Reports* has been a standard by which other texts are measured. My goal in preparing this revised eighth edition has been to keep the best of the earlier editions, to change only that which absolutely had to be changed. Thus the book's old friends will find chapters 5-20, the essence of the book, substantially the same as in previous editions. The first four chapters, the book's setting, are substantially altered. I have added material on audience analysis and adaptation, inserted a short section on the writing process, unified and highlighted the previously scattered sections on headings and visuals by moving them to their own chapters in the front of the book. Jones's various bibliographies have been updated and moved to the end of the book.

I would like to thank Professor Merrill D. Whitburn for first introducing me to the text, my Wm. C. Brown editors Tom Gornick and Susan Soley for their trust and patience, and the various reviewers who through their comments on earlier stages of the project helped me maintain a proper respect for the quality and integrity of Jones's original text. In the tradition of rhetoric that takes *economy* as its catchword, a tradition stretching from George Campbell's *Philosophy of Rhetoric* (1776), through Herbert Spencer's "Philosophy of Style" (1852), through T. A. Rickard's *Technical Writing* (1920), to E. D. Hirsch's *Philosophy of Composition* (1977), it is a pleasure to renew the presence of W. Paul Jones's *Writing Scientific Papers and Reports.*

Preface

[illegible]

Background

1

Preliminaries

By the time your education brings you to this book, you probably have seen many examples of scientific writing. If you have ever read a laboratory report, a physics text, or an environmental impact statement, you are already familiar with scientific writing. Those are only three of countless types of scientific writing you may encounter, or, one day, may need to write.

While you have almost certainly read an abundance of scientific writing, you may never have paused to consider what makes it "scientific." At the beginning of such a course of study as the one in this book, it is useful to take a few minutes to consider where this subject—scientific writing—fits in with other kinds of writing with which you may have become more familiar. The more perspective you have now, at the beginning of the course, the more you will have learned when you reach its end.

This chapter attempts to define scientific writing by comparing and analyzing several examples of what it is and what it is not. While the various kinds of writing shown below cannot be sharply differentiated, and the generalizations drawn about them are at best only approximately valid, the distinctions are real.

WHAT SCIENTIFIC WRITING IS NOT: EXAMPLES

Often the best way to define a complex subject is to state what it is *not*. Thus, we will begin by defining scientific writing by negation (see chapter 5 for more on this technique). Four examples of nonscientific writing are quoted below, each followed by brief critical comments:

1. Emotive Advertising

> A beautiful thing happens with the quiet color of Loving Care. A beautiful thing happens to your hair. Color only the gray without changing your natural hair color. And suddenly your hair will take a turn for the beautiful. You'll be the same girl you were yesterday but you'll feel freer . . . look younger, prettier.
>
> That's what can happen to you with the quiet color of Loving Care.
>
> From a women's magazine

The advertisement presents little information; appeals primarily to love of beauty and youth and to the enjoyment of life; uses emotive words; and is motivated by the desire to sell.

2. Personal, Subjective Writing

> A pigeon peering at me doesn't make me sad or glad or apprehensive or hopeful. With a horse or a cow or a dog it would be different. It would be especially different with a dog. Some dogs peer at me as if I had just gone completely crazy. I can go so far as to say that most dogs peer at me that way. This creates in the consciousness of both me and the dog a feeling of alarm or

> downright terror and legitimately permits me to work into a description of the landscape, in which the dog and myself are figures, a note of emotion.
>
> —James Thurber, "There's an Owl in My Room," *The Thurber Carnival*

The sketch reports no verifiable facts; describes the author's feeling and thoughts; uses emotive words; makes no attempt to influence the reader; generalizes without giving supporting details; and is humorously exaggerated.

3. Slanted Criticism

> That awful power, the public opinion of a nation, is created in America by a horde of ignorant, self-complacent simpletons who failed at ditching and shoemaking and fetched up in journalism on their way to the poorhouse. I am personally acquainted with hundreds of journalists, and the opinion of the majority of them would not be worth tuppence in private, but when they speak in print it is the newspaper that is talking (the pygmy scribe is not visible) and *then* their utterances shake the community like the thunders of prophecy.
>
> —From *Mark Twain's Speeches*

The criticism is satirical and ironic; contains mainly opinion and judgment; uses slanted critical terms; intends to amuse and influence; presents little real information; and is mainly informal and personal.

4. Informative Advertising

> Aluminum has scrap value; it is virtually indestructible. It resists corrosion, will not rust. It can be remelted, re-alloyed, and re-used—economically. And the need for and uses of this strong, lightweight metal multiply yearly. . . .
>
> Our idea is to encourage community groups to sponsor aluminum can collecting drives, and earn money for worthwhile causes and their own needs. As they raise funds, they help keep their streets, parks, and beaches free of litter.
>
> —Advertisement, Reynolds Aluminum

The advertisement is mainly factual and informative; has a concealed motive —to obtain scrap aluminum cheaply; indirectly appeals to civic motives; and is basically persuasive.

As the illustration below shows, we have thus far established boundaries for scientific writing by looking at four examples of what it is *not* (See Figure 1.1). Notice again that these kinds of writing shade into each other gradually.

WHAT SCIENTIFIC WRITING IS: EXAMPLES

The following four examples illustrate the qualities that characterize scientific writing, some of the variety within its boundaries, and how it differs from nonscientific writing.

1. Nontechnical Concrete Explanation

In narrow tubes water does not find its own level, but behaves in an unexpected manner. I have placed in front of the lantern a dish of water colored blue so that you may more easily see it. I shall now dip into the water a very narrow glass pipe, and immediately the water rushes up and stands about half an inch above the general level. The tube inside is wet. The elastic skin of the water is therefore attached to the tube, and goes on pulling up the water until the weight of the water raised above the general level is equal to the force exerted by the skin.

—Charles V. Boys, "Soap Bubbles," from *Soap Bubbles and the Forces Which Mould Them*

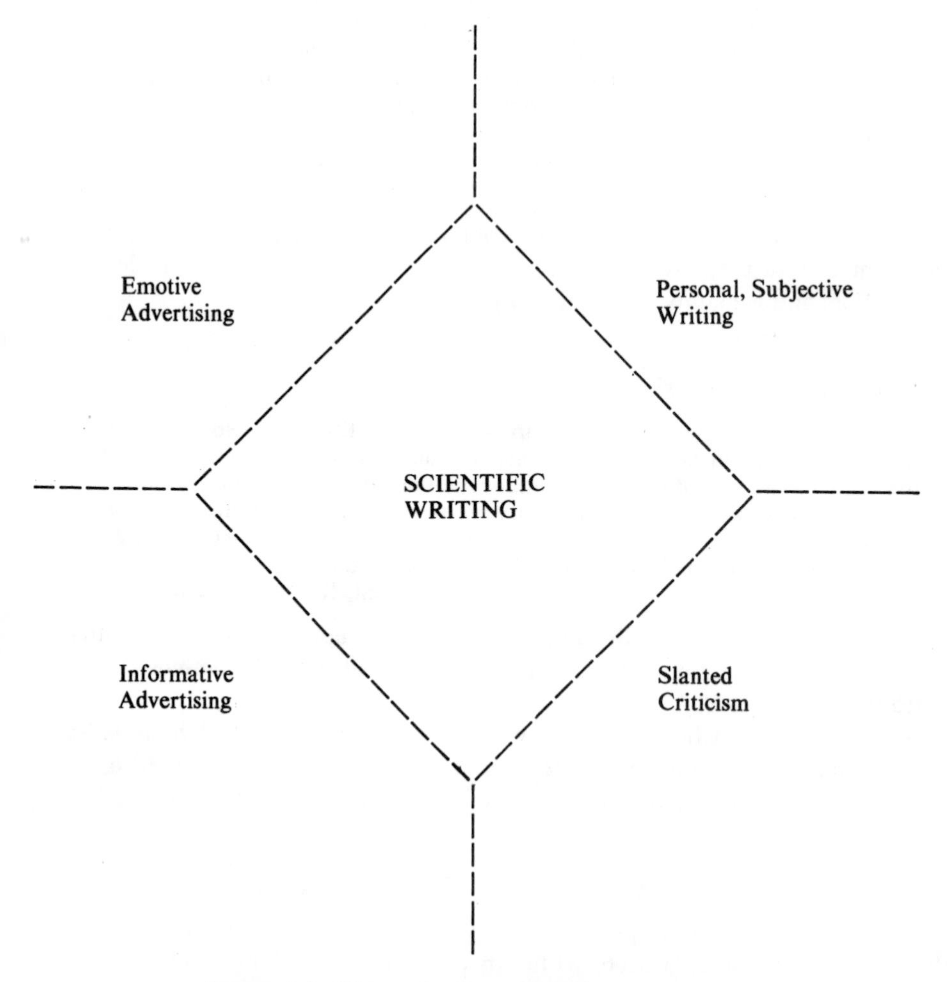

Figure 1.1. What scientific writing is not. Notice the various areas are not sharply separated.

The explanation is intended to inform; bears no evidence of a hidden motive; is nontechnical; begins with a generalization; includes no judgments; and is informal in its use of the first person.

2. Semitechnical Generalized Explanation

> In bright sunshine at midday, enough light falls on a square yard of Earth's surface to yield 8 grams of dry plant matter of average composition in a crop that uses the energy from the light at 4 percent efficiency. The process necessitates an absorption of 11 grams of carbon dioxide. Air contains 0.046 per cent of carbon dioxide by weight, and a cubic yard of air weighs about a kilogram. So 24 cubic yards of air would be needed to supply the 11 grams of carbon dioxide.
>
> —N. W. Pirie, in *Food Resources: Conventional and Novel*

The passage has no other motive than to inform; uses some slightly technical terms; is generalized; makes no judgments; and is formal in tone.

3. Generalized Technical Writing

> How, then, is the signal manipulated to produce the desired permanence? The first step is to route a short pulse being prepared for transmission through a dispersive delay line. The dispersive delay line, which is part of the signal processing equipment, accepts a band of frequencies contained in this pulse, say 25 to 35 MHz, and delays each frequency by a different amount of time to form a much longer RF pulse of, say, 20 microseconds duration. The lengthened pulse will still contain the frequencies between 25 and 35 MHz, the bandwidth needed to make up an RF pulse of 0.1 microsecond duration. This "dispersed" pulse, which is modulated on a microwave carrier transmitted by the radar set, contains 200 times the energy of a 0.1-microsecond "undispersed" pulse transmitted with the same peak power.
>
> —Bell Laboratories *Record*

The passage is informative and formal in tone; assumes a reader familiar with the technical terms used; uses a rhetorical question at the beginning; and is generalized.

4. Generalized Abstract Exposition

> Science is not a system of certain, or well-established statements, nor is it a system which steadily advances toward a state of finality. . . . The old scientific ideal of *episteme*—of absolutely certain, demonstrable knowledge—has proved to be an idol. The demand for scientific objectivity makes it inevitable that every scientific statement must remain *tentative forever*. It may indeed be corroborated, but every corrobation is relative to other statements which, again, are tentative. Only in our subjective experiences of conviction, in our subjective faith, can we be "absolutely certain."
>
> —Karl R. Popper, *The Logic of Scientific Discovery*

The generalizations about science are abstractions with no concrete supporting evidence; are intended to be informative and disinterested; and are nontechnical (the one technical term is defined).

As the following figure shows, we have now defined scientific writing by both exclusion and inclusion (See Figure 1.2). Again, the various areas shade into each other.

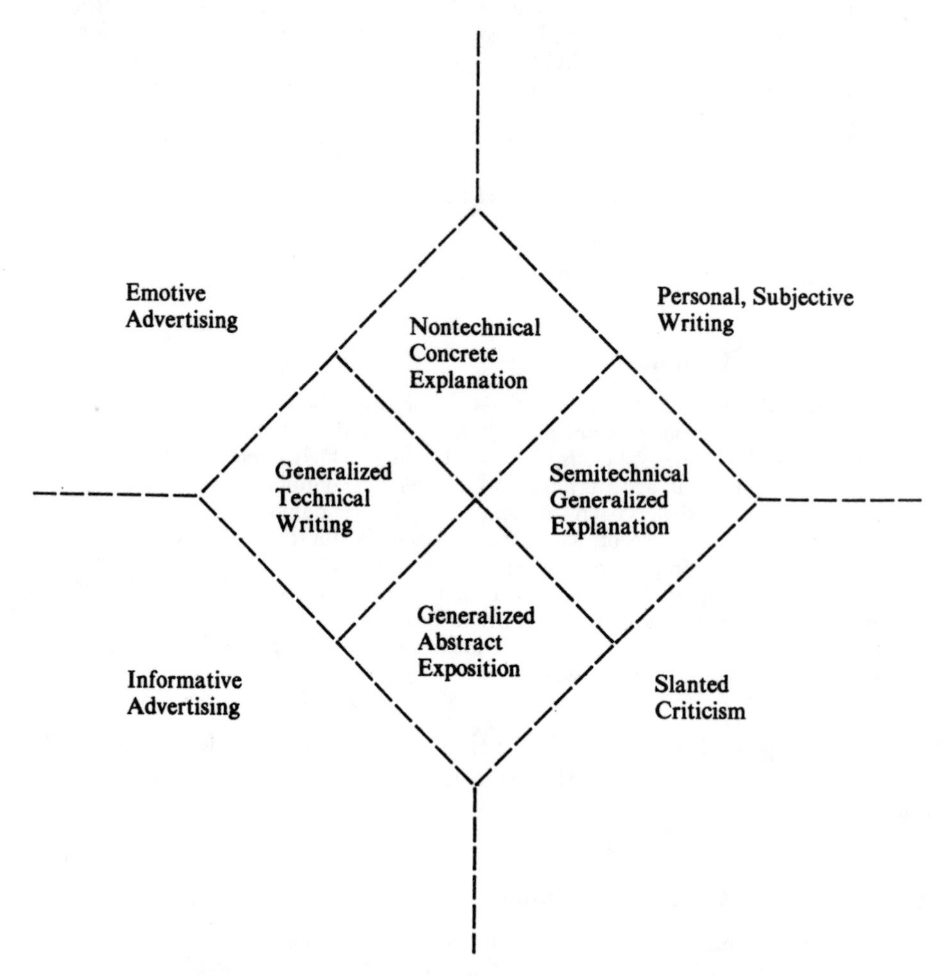

Figure 1.2. What scientific writing is. Again, the various areas are not sharply separated.

SOME GENERALIZATIONS

We can now compare the first four (nonscientific) passages with the next four (scientific) passages, yielding the following generalizations:

Nonscientific writing is likely to be:

Noninformative
Emotive
Subjective
Personal
Entertaining

Scientific writing is likely to be:

Informative
Factual
Objective
Disinterested
Only indirectly persuasive

Both nonscientific and scientific writing may be:

Technical or nontechnical
Abstract or concrete
Specific or general
Formal or informal

Much of our greatest literature is nonscientific. Our culture would be unthinkable without poetry, drama, fiction, history, biography, oratory, and journalism. But while all of these may be factual, often their primary purpose is not to inform. Although the emphasis in this book is upon learning to write scientific papers and reports, it is not intended to imply that such writing is superior. Still there is an excellence which scientific writing can aspire to, an excellence achieved through knowledge and practice, and it is that excellence which this book seeks to help you achieve.

QUALITIES OF GOOD SCIENTIFIC WRITING

The following list summarizes the distinguishing characteristics of good scientific writing:

1. *It presents facts.*

   Science is systematized knowledge. Scientific writing, therefore, presents knowledge in systematic form, or generalizes from systematic knowledge.

2. *It is accurate and truthful.*

   Scientific writers are not content with guesses. Before they write they check and recheck. Regardless of its effects, they tell the truth as they find it.

3. *It is disinterested.*

   The principal motive of its authors is to inform. If they write primarily to persuade, to achieve a selfish purpose, to make money, or to uphold a prejudice, even though they present facts, the result is unscientific.

4. *It is systematically and logically developed.*

   Just as the scientific method is controlled and planned, so scientific writing is controlled and planned.

5. *It is not emotive.*

   Its appeal is to the reason and the understanding, not the feelings. It concerns things in the material world. When it generalizes, it observes the laws of inductive reasoning.

6. *It excludes unsupported opinions.*

   Without facts to support it, an opinion has little place in the method of scientific investigation (except as a working hypothesis).

7. *It is sincere.*

   Since the writer's purpose is to tell the truth as he or she finds it, the writer avoids language that might lead the reader to question his or her sincerity.

8. *It is not argumentative.*

   Its conclusions may vary from the views of others, but they are derived from facts.

9. *It is not directly persuasive.*

   It is concerned with facts and general laws, and with the application of general laws to specific problems. Persuasion is a fine and difficult art, but it is not scientific.

   The effect of good scientific writing may be a change of mind; but the change is due to the use of the scientific method, not to argument, entreaty, or expostulation.

10. *It does not exaggerate.*

    Distortions of facts and unsupported generalizations, any types of exaggeration, are out of place in good scientific writing.

THE TYPES OF SCIENTIFIC WRITING

No classification of the types of scientific writing is complete, because the division lines between types cannot be clearly drawn. The types seldom occur as separate and distinct units. For practical purposes, however, we can set up broad categories which are different from each other in purpose, subject matter, and form. Those categories selected for study here are listed in the titles of chapters 5 through 13 (See the Table of Contents). Besides their use in textbooks, many of these types usually appear only in combination with others. Definition usually combines with classification; descriptions of mechanisms, as well as processes and set of directions, overlap; and analysis is likely to be found in all types of scientific writing.

This book proceeds from types which are usually subordinate parts to those that are usually independent units. For example, any of the types of writing discussed in chapters 5 to 13 may require knowledge and techniques presented in chapters 2, 3, and 4. And the writing of proposals and reports—covered in chapters 14 to 17—may require the use of any or all of the types dealt with in chapters 5 to 13.

Enough types are studied here to give you a good basic knowledge of the general principles governing the composition of all scientific writing. When you know how to write these types, you will be prepared to write any others that may be required in your range of experience.

EXERCISES

A. *Decide which of the following quotations is scientific and which is not*:

1. I the preacher was king over Israel in Jerusalem. And I gave my heart to seek and search out wisdom concerning all things that are done under heaven: this sore travail hath God given to the sons of man to be exercised therewith. I have seen all the works that are done under the sun; and, behold, all is vanity and vexation of spirit. That which is crooked cannot be made straight; and that which is wanting cannot be numbered. I communed with mine own heart, saying, Lo, I am come to great estate, and have gotten more wisdom than all they that have been before me in Jerusalem: yea, my heart had great experience of wisdom and knowledge. And I gave my heart to know wisdom, and to know madness and folly: I perceived that this also is vexation of spirit. For in much wisdom is much grief: and he that increaseth knowledge increaseth sorrow.

—From *Ecclesiastes*

2. John Adams wrote to his wife Abigail from Paris in 1780, "I must study politics and war, that my sons may have liberty to study mathematics and philosophy. My sons ought to study mathematics and philosophy, geography, natural history, naval architecture, navigation, commerce and agriculture in order to give their children a right to study painting, poetry, music, architecture, statuary and porcelain. . . ." Adams' aspiration is still a worthy one.

—Ronald Bergman, Des Moines *Register*, July 28, 1974

3. Some of the activities that are important to us and our sense of being human could, can, and might be programmed; others cannot. To take the extreme case, there simply is no serious sense in which one can talk about a computer program praying or loving. If it continues to be true that to mechanize an activity is precisely to divest it of its *mana*, to cause humans to withdraw from it emotionally, then the impact of these programs, at least culturally, will be to refine our ideas of human intelligence, to cause those ideas to recede, or advance, into the subjective, affective, expressive regions of our nature. If this happens, we might lose interest in the whole issue of whether machines can "outthink" man. . . .

—Fred Hapgood, "Computers Aren't So Smart, After All," *Atlantic*, August 1974

4. *Affect* and *effect* have no senses in common; therefore the tendency to confuse the words must be guarded against closely. As verbs, *affect* (the more common) is used principally in the senses of influence (*how smoking affects health*) and pretense or imitation (*affecting nonchalance to hide fear*), whereas *effect* applies only to accomplishment or execution (*reductions designed to effect economy; means adopted to effect an end*). As nouns, the terms can be kept straight by remembering that *affect* is now confined to psychology.

5. The venerable Professor of Worldly Wisdom, a man verging on eighty but still hale, spoke to me very seriously on this subject in consequence of the few words that I had imprudently let fall in defense of genius. He was one of those who carried most weight in the university, and had the reputation of having done more perhaps than any other living man to suppress any kind of originality.

"It is not our business," he said, "to help students to think for themselves. Surely this is the very last thing which one who wishes them well should encourage them to do. Our duty is to insure that they shall think as we do, or at any rate, as we hold it expedient to say we do."

—Samuel Butler, *Erewhon*

6. This spending of the best part of one's life earning money in order to enjoy a questionable liberty during the least valuable part of it reminds me of the Englishman who went to India to make a fortune first, in order that he might return to England and live the life of a poet. He should have gone up garret at once.

—Henry David Thoreau, *Walden*

7. In mathematics the answer to a problem is nearly always right or simply wrong. In physical and biological sciences, hypotheses may generally be tested and proved correct or incorrect. An administrator can often observe the results of a decision and judge whether he has decided wisely or unwisely. The traditionalist believes that in language, likewise, every construction is either right or wrong. In actuality, however, in language there are no absolutes.

—J. N. Hook and E. G. Mathews, *Modern American Grammar and Usage*

8. If you have ever watched the little corn begin to march across the black lands and then slowly change to big corn and go marching on from the little corn moon of summer to the big corn harvest moon of autumn, then you must have guessed who it is that helps the corn come along. It is the corn fairies. Leave out the corn fairies and there wouldn't be any corn.

All children know this. All boys and girls know that corn is no good unless there are corn fairies.

—Carl Sandburg, "How to Tell Corn Fairies if You See 'em"

9. We can now recognize the very attractive possibility that the left hemisphere of the neo-cortex is suppressed in the dream state, while the right hemisphere—which has an extensive familiarity with signs but only a halting verbal literacy—is functioning well. It may be that the left hemisphere is not entirely turned off at night but instead is performing tasks that make it inaccessible to consciousness: it is busily engaged in data dumping from the short-term memory buffer, determining what should survive into long-term storage.

—Carl Sagan, *The Dragons of Eden*

10. Respiratory pigments are found in many kinds of animals, and all of the pigments have the same constitution: a large protein molecule with one or more prosthetic groups. In vertebrates the respiratory pigment is hemoglobin, the prosthetic group of which is heme, an iron-containing porphyrin.

Hemoglobin, as you well know, has a dark purple color in the absence of oxygen and becomes bright scarlet when allowed to take up oxygen.

—Todd-Sanford, *Clinical Diagnosis*

B. *Bring in examples of both scientific and nonscientific writing selected from your reading, with your analyses of their characteristics.*

C. *Write paragraphs exemplifying the following types of prose:*

1. Scientific writing, nontechnical, concrete
2. Scientific writing, technical, concrete
3. Scientific writing, general, abstract
4. Nonscientific writing, persuasive, emotive
5. Nonscientific writing, unsupported judgments
6. Nonscientific writing, subjective, personal

D. *Write sentences in which you use five of the following with different meanings and connotations.*

For example, *pig*:

1. The *pigs* in the pen are large enough to be marketed.
2. "We don't like *pigs* in our part of town," yelled the gang leader at the police car.
3. Crude iron cast in blocks is called *pig* iron.

square
rich
lid
fix
jerk
thing
tap
bull
short
function
loot
bloody
cool
straight
home
awful
date
mad
bust
lay
line

2

The Method of Science and The Process of Writing

To write a good scientific paper or report you need something more than the mere knowledge of facts, though knowledge of your subject is indispensable. How people get the notion that mastery of facts is all it takes to write a paper presenting facts is difficult to understand. Yet it is a common misconception.

Writing is like any other skill: to acquire it requires knowledge and practice. At its best it is an art comparable to playing a musical instrument or a game; no one would expect to play even a ukelele well or to equal par in golf without first learning how and then practicing. Yet scientists, managers, engineers, technicians, and others who have never written anything except laboratory notes or school essays are indignant when their report or paper is refused for publication by the editor of a technical journal, or when their report is deemed unacceptable by their superior.

REASONS FOR STUDYING SCIENTIFIC WRITING

Dr. Morris Fishbein, in chapter 1 of *Medical Writing,* 1938 edition, listed the following among the reasons—other than lack of space—for the rejection of papers by the *Journal of the American Medical Association.*

1. The facts stated in the paper are already known.
2. The author does not know the facts about his subject.
3. The conclusions are not justified by the evidence given.
4. The paper is not based upon firsthand observation.
5. The paper is intended for oral delivery and not for publication.
6. The material of the paper is poorly organized.
7. The style is not adapted to the readers of the *Journal.*
8. The material is not sufficiently condensed.
9. The writing is ungrammatical, verbose, obscure.
10. The author tries to dress up his facts in fine, fancy style.
11. The same point of view is not retained throughout.

Perhap the biggest problem with scientific writing today is the writer who assumes that the reader necessarily shares the writer's knowledge of or interest in the subject. The use of this book is designed to help you avoid this fault and the faults listed above. Although study and practice may not make you a Thomas Henry Huxley or a William Beebe—or an Isaac Asimov, or a Carl Sagan, they will at least enable you to write a good scientific paper or report.

The clearest way to explain the difference between knowledge of a subject and written presentation of that knowledge may well be to compare the ways in which knowledge and written presentation are accomplished. Any person engaged in scientific writing will have used some variation of the scientific method to arrive at knowledge; any person presenting that knowledge will use some variation of the writing process. The following pages present a comparison of the scientific method and one version of the writing process.

THE METHOD OF SCIENCE: TEN STEPS

"The method of scientific investigation," wrote Thomas Henry Huxley, "is nothing but the expression of the necessary mode of working of the human mind." The ordinary person often follows the method without realizing it, hastily and carelessly; the scientist proceeds consciously from step to step, slowly and cautiously. The scientific method is a strictly controlled procedure beginning with observation and experiment and ending with general statements.

Although there are many differences in the *technological* methods used in the different areas of science, the *logical* methods are about the same in all areas. Actual research seldom follows a rigidly fixed procedure. Unexpected events, accidents, and disturbances interrupt and distract. A piece of research is an event in history, and history allegedly never repeats itself. To say, then, that the scientific method is a completely standardized procedure would be incorrect. Ideally, however, research will follow a certain well-established sequence of steps. Basically we can distinguish the ten steps listed and briefly explained below. Again let us emphasize that *actually* the procedure frequently varies.

1. Defining the Problem

Scientists must have an objective other than the vague ambition to find something new. They must define a problem as specifically as existing knowledge will permit. What is the cause of dew? Is yellow fever contracted from the bite of a mosquito? What type of pavement is most efficient for heavy-duty truck-traffic?

2. Summarizing Work Already Done

Efficient workers make sure that their problem has not already been solved, and they find out what other work has already been done upon it. They search the literature of the field and carefully record any evidence bearing upon the problem.

3. Comparing Similar Phenomena

No matter what the problem is, there probably have been others like it. The solution of one problem may be useful in finding the solution of a similar problem. The worker who seeks the cause of a disease would naturally try to find out what causes of similar diseases are known and what methods were used in finding those causes.

4. Forming a Hypothesis

With the problem clearly stated and with a record of all previous work on the problem, the next step is to formulate a working hypothesis, a tentative answer to the problem. Perhaps this is the most important step, calling for the greatest exercise of imagination and intuition.[1] Discovery of the great laws of nature has not been the result of following a mechanical routine. Thus Torricelli's concept of atmospheric pressure—that the earth is surrounded by a "sea of air" that has weight, and that objects submerged in it would be subject to air pressure just as objects below the surface of the sea are subjected to water pressure—was a flash of genius completely opposed to the established concept of a "full" universe in which a vacuum is impossible.

5. Testing the Hypothesis

The next step is to establish, by observation and formal experiment, the falsity or the truth of the hypothesis. Thus Torricelli devised and performed the classic experiment with a tube of mercury inverted in a dish of mercury, produced a vacuum, and demonstrated that air has pressure. In such tests scientists guard especially against a prejudice in favor of a hypothesis that might result in overlooking or underemphasizing negative evidence. Especially do they try to disprove their hypotheses.

In this and the following steps the methods and the apparatus and the instruments used may be highly technical and will vary greatly in the different fields of science. If possible, scientists devise and carry out rigidly controlled experiments, with only one of the possible factors concerned being varied at a time. They record the results as supporting the hypothesis.

6. Extending or Modifying the Hypothesis

As evidence accumulates, scientists may achieve any one of several possible results. They may verify the working hypothesis; they may disprove it; they may extend its original scope; they may modify or restrict it. Eventually they will discard it, retain it, or restate it to correspond with the data obtained.

Carrying out steps 5 and 6 may require innumerable experiments involving hundreds of people and extending over many years.

7. Testing the Extended or Modified Hypothesis

Whenever scientists are forced to alter their hypothesis to correspond with newly discovered facts, they must repeat old experiments or devise and carry out new ones to test the validity of the new statement.

8. Publishing the Results

Not until a final statement of the hypothesis has been exhaustively tested and established will the results be published. Of course, from time to time progress reports may be issued, especially for the benefit of colleagues. But publishing anything positive and final is postponed until the possibilities for further observation and experiment have been exhausted.

9. Submitting the Final Hypothesis for Verification

Steps 8 and 9 may be simultaneous or their order may be reversed. Before risking publication, scientists may privately submit the results for others to criticize. At any rate, when they publish the results, they expect that others will test the hypothesis, perhaps repeat the experiments and devise new experiments. If the original work was as systematic and exhaustive as it should have been, other scientists will merely succeed in strengthening the published results.

10. Establishing a Theory

Only when this process of checking and rechecking has been completed to the satisfaction of competent workers in all parts of the field does the hypothesis acquire the dignity of a theory.

* * * * *

If "the method of science" were as simple and invariable as this list of steps would indicate, we might well wonder why it was not discovered and followed centuries earlier. But actually this analysis much oversimplifies the procedure. As James B. Conant says in *Science and Common Sense,*

> The stumbling way in which the ablest of the scientists in every generation have had to fight through thickets of erroneous observations, misleading generalizations, inadequate formulations, and unconscious prejudice is rarely appreciated by those who obtain their scientific knowledge from textbooks. . . . To attempt to formulate in one set of logical rules the ways in which mathematicians, historians, archeologists, philologists, biologists, and physical scientists have made progress would be to ignore all the vitality in these varied undertakings.

Such an analysis should be supplemented by studies of procedures actually followed by scientists.

THE PROCESS OF WRITING: FOUR STAGES

Just as the method of science as it is actually used always differs at least slightly from any textbook presentation of it, so the process of writing inevitably differs from individual to individual and from situation to situation. Let us look briefly at one widely accepted view of writing. After this, we will compare the method of science and the process of writing. As we shall see, there are interesting differences and similarities.

Many writers describe their composing processes as having stages similar to the following: *planning, writing, reflecting,* and *revising*. While an individual writer's names for these stages may vary somewhat, and the characteristics of each stage are certainly not the same from person to person or from situation to situation, the following discussion captures the experience of many writers.

Planning

As with any other complex activity, writing requires planning. While the precise nature of your planning will vary from task to task, the following three activities are essential.

Gather Materials. Most writers try to assemble everything they need in the way of data and information before they begin to write. If your project is so large that you cannot mentally "grasp" it all at once, perhaps you should break it down into smaller, more manageable sections when you write it down. The more you know about what you want to say before you begin, the easier your writing task will be.

Determine Purpose and Form. In scientific writing, your purpose and form are often dictated by the situation. If the metallurgical laboratory you work for directs you to prepare a report on the analysis of a sample, there will usually be a prescribed format for you to follow, and the purpose of your report—perhaps to determine why a structural member failed—will most often have been explained to you. Any additional effort you have to expend to determine the purpose of your writing is usually worth it. Devoting a little time to finding out about the proper form, even to the point of examining a similar piece written by someone else, can save you a great deal of time redoing your report later on.

Determine Audience. As the next two chapters explain, different audiences make diverse demands of writers. You need to know as much as you can about your readers, especially who they are, whether they know (or care) very much or very little about your subject, and what they expect from your report. Even if you have never seen your readers, you can use what you know about them to draw their pictures in your mind and to make some important decisions about what this or that kind of reader will want, need, or expect from you.

Writing

For most of us who write professionally, the situation usually forces us to write before we are fully ready. Inexperienced writers often compound this problem by underestimating the time writing requires, calculating only the time needed to put the words on paper, but omitting the time their ideas need to grow into words. In scientific writing, because of the complexity of content and the frequent inflexibility of deadlines, it is best to begin with as much already on paper as possible and to plan on at least two drafts to achieve a finished product.

Get a Running Start. Most inexperienced writers have trouble when they begin with a blank page. The best way around this is to avoid starting with a blank page by having stacks of notes, pages of records, and in general an abundance of words to choose from before you begin. If your writing task is such that you have access to written records, you have a mine of words to work from. If not, the instant you receive your assignment you need to begin jotting down ideas, words, and phrases as they come to your mind. Never trust your own memory to recall that good opening line you thought of while at lunch; let the note pad do the remembering for you. If you have an abundance of materials in written form to begin with, you should be able to get a good, running start.

Plan For Two Drafts. Many writers have a hard time getting their thinking on the subject exactly straight until they have already written their thoughts down once. "How can I know what I think until I write it down?" is a common expression for many of us. Give yourself the luxury of time for a second draft—a margin for error, if you like—by assigning yourself a work schedule and adhering to it. Often the difference between good writing and poor writing is not the different skill levels of the writers, but their different levels of discipline and determination. Plan your time to allow for two drafts, at least, and if the day comes when you only have time for one, the experience you have gained in the past by writing two drafts will make you better able to deal with the one-draft situation.

Reflecting

Perhaps the worst thing you can do as a writer is to finish a piece and turn it in immediately. You may often be forced to, but such situations do not make for the best writing. The technique which professionals use most, and amateurs least, is allowing time to reflect on what they write. The passage of time allows one to get a fresh perspective on those words that seemed so right earlier (and, often, so wrong later). Once again, scientific writers often have too little time to begin with; therefore they must develop other ways of gaining perspective.

Time. How long does it take between finishing a draft and seeing it through fresh eyes? At least overnight, and at most a week, a month, six months, or a year. When you have the luxury of time, use it to your advantage. When you do not have time, try the following other techniques for acquiring a fresh perspective.

Perspective. In professional life, the most commonly used tactic to gain a different outlook on a paper or a report is usually to hand it over to a co-worker for feedback. Having a friend in the next office who is a good "reader" has made many a successful writer. Of course, most colleges do not allow students to do this, or if it is allowed, it is only permitted with tightly drawn guidelines (check with your professor).

There are other ways to gain a fresh perspective. One of the most common is to go from a handwritten draft to a typed one; somehow the altered appearance of the typed copy allows one to see the content and needed changes more clearly. Another method is to write first thing in the morning and reflect on it in the afternoon. Or you can try "talking it out" to someone else, whose comments and opinions may help you recast your own image of your writing.

Finally, you need to ask yourself these questions: Does your paper really fulfill its purpose? Does it suit the specific audience you aimed at? Does it convey its message accurately? Taking the time to give these questions serious and thoughtful answers may well be the best technique available to students for gaining a fresh perspective on their writing.

Revising

Doing a good job of revising should be the critical point for producing good writing, but too often it receives the least attention of any of these four stages. Careful and thorough revision can turn the densest, most pedestrian style into a model of clarity. There are three keys to such revision: looking for order, trying to sense possible problems, and being as demanding of yourself and your writing as the situation allows.

Look For Order. At every stage of your approach to any writing task you should look for order. Try to see beneath the structures and arrangements that may have characterized your first thoughts about your material to select the patterns that can make it work best for your specific purpose and audience. You need to recognize the order which your content suggests, but you also need to realize the extent to which that order is susceptible to variation. And this search for order should take place at every level of the text, from word-to-word to chapter-to-chapter.

Sense Possible Problems. Look for instances where the way you want your text to read and the way it does read conflict. You may be able to detect these places by reading the text aloud; some people even tape record and play back their papers so they can devote all of their attention to listening. Be scrupulous about whether your evidence supports your claims, about how much effort you ask your reader to spend to understand you—and whether your reader will spend it. Good writers remember their own version of Murphy's Law: "Anything that can be misunderstood probably will be."

Be Demanding. Your finished work will almost certainly carry your name on it, and it may be the basis for whatever reputation you have in your colleagues' eyes. Do not give in to the temptation of believing that the excellence of your methodology and the soundness of your results will outweigh the poverty of your prose. If no one else can or will understand your writing, or if the readers misinterpret or skim it because it is so poorly written, what have you gained?

* * * * *

The last and most striking characteristic of the writing process is its *recursiveness*; that is, at every stage it can loop back on itself. Often the scientific method is at its best when it is the most *linear*—strictly controlled, slowly and cautiously proceeding from step to step. Writing often goes best when its process is the least linear. In the middle of one sentence a writer may reread the previous one for fresh inspiration. While gathering information for one section you may restructure the previous one, or outline the next one. Or in the middle of typing what you hoped would be the final draft you may have a "Eureka" moment and realize how it really should be done, rip the paper out of the typewriter, and start cutting and taping your old paragraphs into a new structure. Experienced writers take advantage of this recursiveness and use it fully; they keep their options open to go back and forth through these stages as their needs and judgments dictate. The following figure illustrates this recursive quality (See Figure 2.1).

METHOD AND PROCESS COMPARED

The following table compares the method of science and the process of writing as described above:

TABLE 2.1
Comparison of the method of science and the process of writing

Method of Science (linear structure)	Process of Writing (recursive structure)
1. Defining the problem	Planning
2. Summarizing work already done	Gather materials Determine purpose and form
3. Comparing similar phenomena	Determine audience
4. Forming a hypothesis	Writing
5. Testing the hypothesis	Get a running start Plan for two drafts
6. Extending or modifying the hypothesis	Reflecting
7. Testing the extended or modified hypothesis	Time Perspective
8. Publishing the results	Revising
9. Submitting the final hypothesis for verification	Look for order Sense possible problems
10. Establishing a theory	Be demanding

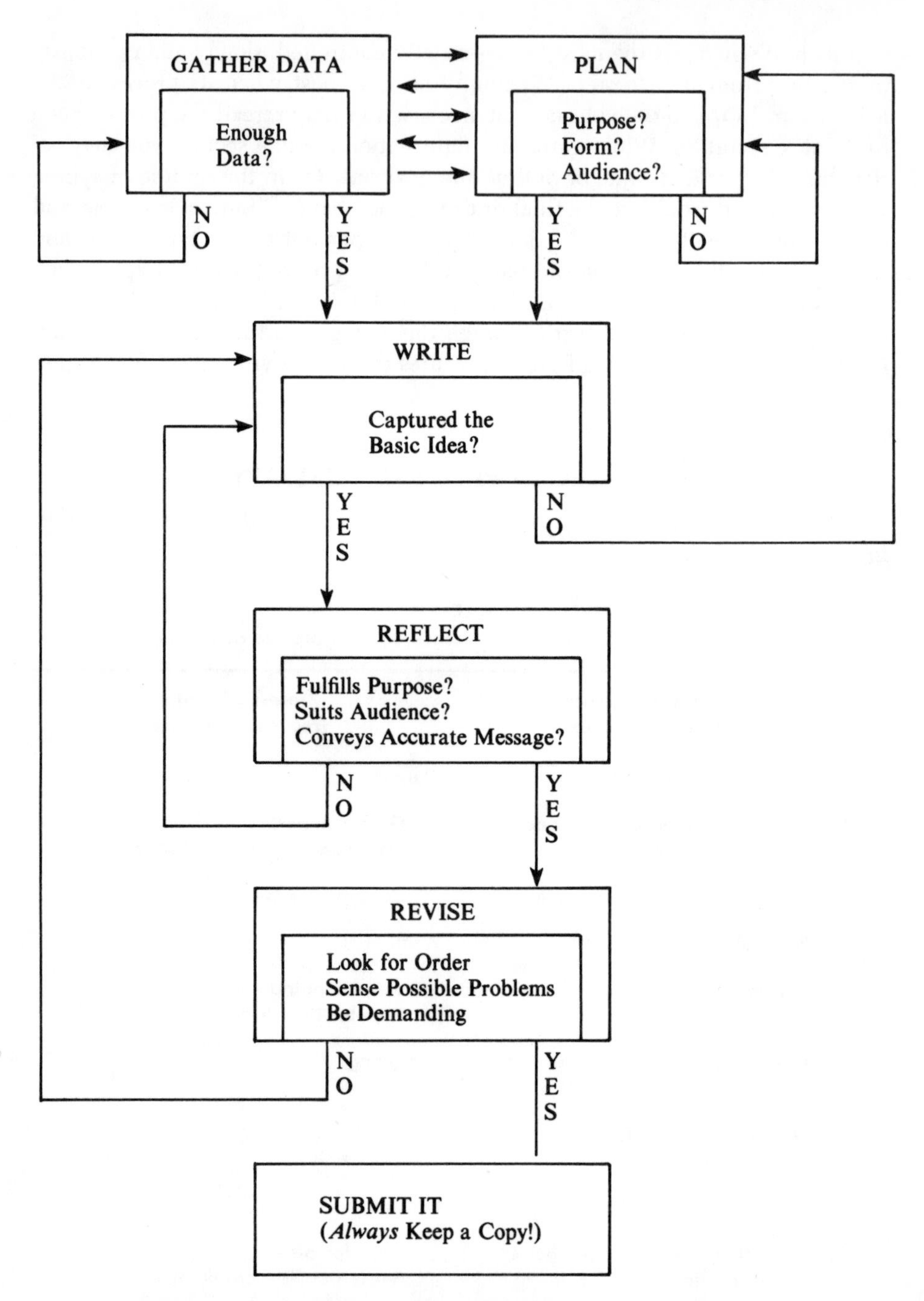

Figure 2.1. The process of writing. Flow chart is designed to indicate clearly the recursiveness among the various stages.

As examination of Figure 2.1 and Table 2.1 shows, there are interesting similarities and significant differences between the method and the process. Although there are certainly exceptions to the linearity of the scientific method, it is characteristically orderly and incremental—each step builds on the previous one. The composing process is at times linear, but characteristically recursive, and occasionally nearly chaotic, with subsequent steps at times eliminating previous ones.

It is tempting, however, to put each of the two into a four part structure (as suggested by the vertical spacing in Table 2.1): "Planning" corresponds to roughly steps 1-3, "Writing" with 4-5, "Reflecting" matches 6-7, and "Revising" matches 8-10. These similarities show that the method of science and the process of writing are similar in what we might call their rhythms, that is, doing background work, making an initial advance, refining it, and making it final. But once we come to the specific characteristics of each step, the similarities, for the most part, cease.

While the process of writing and the method of science are in their particular details separate skills, one other similarity is important here. Both are complex intellectual skills, and both must be learned.

THE NEED FOR PLANNING

Good planning combines following the scientific method with writing a good report. Before you write any scientific paper or report, you need to plan, and the time to begin that planning is while you are still gathering materials. Authors of personal letters may, if they like, follow the stream-of-consciousness method and put their ideas down in any order. Their recipients may be grateful for any message and may not mind the lack of a beginning, a middle, and an end. Such stream-of-consciousness writing may be said to be writer-centered; that is, it shows the order in which the ideas came to the writer, with no concern for the reader's needs.

The structure of a good scientific paper or report, however, needs to be reader-centered: it needs to respond to the needs and interests of the reader. What is there about your report that will interest the reader, and how can you draw his or her attention to it? What is it that will cause the reader problems, and how can you minimize that characteristic? Most important, can your reader *see* the structure of your paper or report? Does that structure reflect the order your material first assumed in your mind when you collected your data (a writer-centered structure), or have you re-organized your material into the structure through which your reader can best understand you (a reader-centered structure)? To structure your scientific paper or report this second way,—according to the reader's needs and interests,—requires planning. And the longer your report is, or the more complex its content, the more structural planning you need to do.

Too often writers approach their tasks without sufficient planning. Their minds may be full of their subjects so that they are aware of every detail, but their task in writing is not only to impress their readers with the extent of their knowledge, but also to present that knowledge in an orderly fashion. What should they write first, second, and so on? The planning should not be left to chance.

Some writers have an intuitive sense of form. Without conscious thought they may become aware of the divisions and subdivisions of their material and put things in logical order. Others write themselves into their subject, as it were, and later juggle the pieces into proper arrangement. Some great novelists—Charles Dickens and Thomas Wolfe come to mind—never knew when they started a story how it was going to end. But they were geniuses.

You, the average person facing a writing task, will do well to proceed differently. First of all, you need to know your subject. How you get this knowledge is not the major concern of this book. You may spend interminable days and nights in a laboratory; or you may, as Charles Darwin did, range the world for years observing and collecting. You may question other people to find your facts; or you may, like Albert Einstein, merely sit and think. But eventually you will need to communicate your knowledge to others. Then you may appreciate some help in going about your task systematically.

GENERAL PRINCIPLES OF ORGANIZATION

The general principles which should govern your planning will usually be evident in the nature of your writing task; either your material or your evidence will often dictate the organizational principles you will use. You should be aware of the more obvious ones.[2]

Chronological Arrangement

Whenever the collection of material takes time, as in many experiments, tests, jobs of work, and field trips, the chronological arrangement may be the best. To relate events in the order in which they happened or were observed may be the simplest and most effective way of arranging them. The course of an experiment in the growing of a crop of corn, from selecting and testing seed to harvesting, you could so report.

Space Arrangement

Whenever a subject has to do with an area, whether small or large, dividing the area into its parts may be called for. The description of a new oil field might be best presented as subdivisions of the whole field. The description of the layout of a factory, store, or farm (see the chapter on Organizations) would probably be according to space. The same principle may be applied in descriptions of mechanisms, where the parts are taken up in turn.

Partition, or Division

Where any whole may be broken down into its parts, the principle of partition, or division, will be a likely choice. We understand the whole by studying its parts and their functions and relationships. A report on the composition of any chemical compound would be likely to proceed by this principle.

Causes and Effects

After study has revealed the causes or effects of any phenomenon, the treatment of these in their order of importance would be logical. The arrangement might proceed from most important to least important, or the reverse. An engineer might so organize a report on the failure of a bridge, the breaking up of a highway surface, or a failure of electric power. The overflowing of a river, the failure of a business, the raising of taxes, all might be analyzed from the points of view of either their causes or their effects, or both.

Classification

Where many items or details are to be presented, whether they are concrete or abstract, their classification into groups according to their similarities and differences is often a basic principle of organization. A report on the fish in Iowa streams and lakes, or the soils of Iowa farm lands, or the trees in Iowa woodlots would probably be so organized.

Citation of Examples

Some subjects may call for examples of a phenomenon. A study of the pollution of water in a river, of the dying out of passenger traffic on our railroads, of the causes and cures for inflation—a host of subjects can be most effectively presented by citing examples. The method is perhaps less scientific than some others because the examples may be incomplete or atypical, or chosen to support a prejudiced main idea. They must be chosen with great care if the result is to be good scientific writing.

* * * * *

The headings above indicate some of the main principles for organizing the material in expository papers of some length. Of course you are not restricted in any paper to the use of a single principle. The parts of a paper using a chronological or space principle of arrangement may be expanded by use of partition or division, by a list of causes and effects, by a classification of parts, or by an example or series of examples. But the basic principle you should adhere to throughout, and subordinate the use of other principles.

Construction of an Outline

At the beginning an outline is merely a tentative plan. On a scratch pad put down a list of all the items you can think of that are pertinent to your subject. The result will be a haphazard list. Don't worry at this stage about form or arrangement. Study the example of a typical procedure in the chapter on Analysis.

When you can think of nothing else to add, examine the items carefully and make a new list, grouping those which logically belong together. Then work with each group in turn to distinguish main items from subordinate ones. The result will be the beginning of a working outline which can be your guide as you write.

But don't let the working outline dominate your thinking. It is not a blueprint to which you are obliged to conform. As you write, you may recall material you haven't provided a place for, or you may think of a better arrangement. If necessary, modify your plan as you proceed. Eventually it can be made into a formal outline that may become a table of contents whose items correspond with the headings and subheadings of your paper.

Forms of Outline

Three forms of outline are in common use: the topical form, the sentence form, and the combination form. Which form you use depends on the purpose you want the outline to serve. A topical outline may be sufficient to help you plan your report, but if you want to explain that plan to someone else the greater detail of the sentence outline may be more desirable.

The Topical Outline

The topical outline consists of a series of items, grammatically incomplete constructions, designating what the various sections of the exposition are about. It will always contain at least two main heads, probably more, and probably subheads; for long and complicated analyses, down to the fourth or fifth levels of subordination. The topics do not summarize what you say about them; they merely indicate what the sections are about. For one's own guidance in planning, this form is best.

Illustrations of topical outlines may be found in this book at the beginnings of all chapters, where the subject matter and relative importance of sections are indicated. Other examples are in the chapter on Classification and the tables of contents of long reports, chapter 16.

The Sentence Outline

The sentence outline summarizes each section in a complete sentence. For the reader unfamiliar with the subject matter it is more informative than the topical outline, because it does more than indicate the subject of a section: it summarizes what is said about that subject. The arrangement is substantially the same for both forms. Illustrations may be found in the chapter on Comparison, where a generalized topical outline is followed by a sentence outline.

The Combination Form

In the combination form the main sections are summarized in complete predications, but the subsections are left as topics. This kind of outline is often easier to make than a sentence outline, and may avoid unnecessary repetition.

A typical example is in the chapter on Abstracts, "Example of Procedure in Preparing an Abstract."

Use of Letters and Numbers

The use of letters and numbers in an outline is optional. When the outline is long, the symbols may help the reader to keep the relationships of parts clear. When the outline is short, they may be of little use. What is said on page 45 concerning the use of letters and symbols with headings applies as well to their use in outlines.

Parallelism of Structure

In outlines to be used by persons other than the writer, it is important that all items of the same series be grammatically parallel. See the section on Parallelism in chapter 20. Whether you observe the principle in a working outline for your own use is unimportant.

* * * * *

To know your subject well is the first requirement in scientific writing. To have a good plan of organization, either in your head or in outline form on paper, is the second. But there is a third no less important: to use the information and devices available to you for making the plan clear to your reader. That third requirement is the subject of the next two chapters.

EXERCISES

A. *Read an account of some great scientific investigation and report on the extent to which the procedure corresponded to that outlined in this chapter. Suggested for study are the investigations of such people as Galileo, Newton, Faraday, Franklin, Darwin, Harvey, Boyle, Dalton, Maxwell, Curie, Lavoisier, Pasteur, Koch, Bell, Freud, Michelson, Mead, Goodall, and Watson.*

B. *Work together with several of your classmates to design a container that will enable a fresh egg to withstand a ten foot drop onto concrete. Write a brief summary of the process by which you worked, and compare that process with the scientific method as outlined in this chapter.*

C. *Choose a skill that you do well, such as bicycling, fishing, weaving, or back-packing, and make a topical outline, a sentence outline, and a combination outline for a short "how to" pamphlet on the subject, designed to introduce a teenaged beginner to that skill.*

D. *Make a combination outline of your own writing process. Be as detailed as you can.*

Notes

1. "The great working hypotheses in the past have often originated in the minds of the pioneers as a result of mental processes which can best be described by such words as 'inspired guess,' 'intuitive hunch,' or 'brilliant flash of imagination.' Rarely if ever do they seem to have been the product of a careful examination of all the facts and a logical analysis of various ways of formulating a new principle."—James B. Conant, *Science and Common Sense,* page 48.

2. Refer to the section on Expanded Definitions in the chapter on Definition and also to the chapter on Analysis for further explanation and illustration of the use of these principles.

3

Remember the Reader I: Audience Analysis and Adaptation

T. A. Rickard, author of one of the century's first textbooks on scientific and technical writing, once declared: "This is the first great principle of writing—economy of mental effort on the part of the reader." When a writer assumes the reader's knowledge of the subject or interest in the subject mirrors the writer's own, it usually results in extra effort on the part of the reader. A specialist writing for other specialists in the same field often feels justified in making the audience work to understand the text, under the assumption that since the audience mirrors the author's knowledge and interest concerning the subject, they are willing to put as much work into reading the text as the author put into writing it. Unfortunately, that is often a mistaken assumption. A skillful writer will take the time and trouble to write for a *real* audience, making a realistic appraisal of their knowledge, their interest, and especially their willingness to expend effort in their reading. A skillful writer remembers the reader.

THE NEED FOR AUDIENCE ANALYSIS AND ADAPTATION

In every field the number of situations requiring one to write for an audience which is *not* composed of specialists in that field is on the increase. Architects write for bank officers, engineers write for Congressional committees, chemists write for management, and everyone writes for government bureaucrats. Because the number of situations requiring one to write for non-specialist audiences is increasing, training in that kind of writing is essential. The principles of audience analysis and adaptation you will learn from writing for a non-specialist audience will also make the writing you do for other audiences clearer. Training in writing for a variety of kinds of real audiences, especially non-specialists, can best be furnished under two headings: audience analysis and audience adaptation.

AUDIENCE ANALYSIS

Remembering the reader means writing for a specific, real audience. Two factors about the real audience need particular consideration: What is your audience's *knowledge* of your subject? And what is your audience's *interest in* your subject? The easiest way to generalize about answers to these questions is to classify audiences under some broad scheme. We can distinguish four kinds of audiences: expert, executive, technician (or operator), and layman.[1]

Expert

An expert audience is composed of people with substantial knowledge of and interest in one's subject. Often such people hold advanced academic degrees, or positions of great responsibility, or both, in the particular field being discussed.

Examples would be a research chemist writing a paper for other research chemists, a physicist preparing a paper for presentation at a convention of the physicists' professional society, or a physician writing for the *Journal of the American Medical Association.* In each case, the writing may well be nearly unintelligible to laymen, and difficult perhaps even for other experts. Few concessions are made to spare the reader "mental effort," and the writer often takes the reader's interest in the subject for granted. While many experts in every field wish that their fellow experts would make more concessions to their readers, such is unfortunately not very often the case.

Executive

An executive audience is one composed of decision-makers. Specifically, one is writing for an executive audience whenever one writes for an audience which has decision-making power over one's life. This may be a person applying for a job, a scientist writing to his or her superior, anyone applying for a grant, or a corporate vice-president preparing the year-end report for the board of directors. In such cases, one needs to keep in mind that the audience's interest in the subject may well need to be created (or at least refreshed in memory) by the author. And the audience will nearly always particularly appreciate brevity and clarity. It should also be noted that executive audiences are almost never "pure" executive; usually the executive will either be also an expert (an executive/expert audience) or a layman (an executive/layman audience). It is almost never sufficient only to know that one's audience is executive. Is it executive/expert, or executive/layman?

Technician or Operator

The technician or operator audience is composed of the people who operate the machines, program the computers, service the airplanes, repair the cars, or pour the concrete. In the Army they may be the non-commissioned officers, and in the university they may be the secretaries or the teaching assistants. The technician or operator may be a high school graduate employed at the lower level to work with hands more than head, or the technician or operator may in effect be an apprentice expert, such as an auto repairman studying to be a mechanical engineer, or a teaching assistant working on a Ph.D. degree. Both kinds of technician or operator share crucial qualities: they want clarity, brevity, and results—a minimum of theory and a maximum of practical application. While they may be interested in your subject, it is interest that needs to be specifically addressed in practical terms. Something on the order of "Read this manual carefully and follow it step-by-step to insure your preventive maintenance really is effective" is the kind of approach needed for technicians or operators. More so than for any other audience, a writer addressing technicians or operators needs to keep both feet on the ground.

Layman

A layman audience is composed of people who have no more than a superficial knowledge of the subject—if that much—and no intrinsic interest in it. A corporate executive writing a speech for the local PTA, an astronomer writing for the daily newspaper, a teacher writing for undergraduate students, and an architect or engineer writing for prospective home-buyers, all are examples of writing for laymen. No other audience requires the writer to make so many different kinds of concessions to the reader, concessions in terms of the kind of vocabulary, the familiarity with concepts, the kind and amount of detail. A layman must be shown why to read. A layman must be led through the text. And a layman wants to be met more than halfway in terms of the content, style, and organization of the written statement.

* * * * *

Each kind of audience has its own characteristics and its own requirements. Recognizing the audience's type—executive, expert, technician (or operator), and layman (or layman/executive)—is only the first step in remembering the reader. The second step is equally important: a successful writer must not only *analyze* the audience, but also *adapt to* that audience.

AUDIENCE ADAPTATION

Writers whose work succeeds know that different kinds of audiences pose different types of problems. Here are just a few of the areas one must consider when adapting to a specific, real audience.

Content—Is the content too deep, too shallow, too long, too short, exactly relevant, or irrelevant?

Concepts—Are the concepts already familiar to the reader, and if not, are they sufficiently necessary to the paper to justify the time and effort explanation will take?

Vocabulary—Are these words clearly familiar to this audience? If they are marginally unfamiliar, can they be briefly and clearly explained? If they are totally unfamiliar, can they be replaced by easier words, or are they necessary at all?

Organization—Does this audience want to be shown procedures first, or results first? Which is the most important to them?

Method of Development—Will this audience understand causal analysis, or is personal narration better? Is a brief history of the problem the best way to prepare for this solution? Is comparison and contrast of other solutions the best way to justify this course of action?

Visuals—Are charts and graphs easily understood by this audience? Do sketches require complete detail? Would flow-charts help get the process clearer?

As you may have guessed, audience adaptation can affect every aspect of a piece of writing, its content, organization, and style. Because of the variety of choices under audience adaptation, let us look in detail at the process and methods of adapting to only one kind of audience—the layman audience. Since that audience requires the most adaptation, a discussion of methods of adapting to lay audiences can serve as a guide for principles of adaptation to any kind of audience. The same principles may be said to apply, but in different ways and to different degrees, depending on the type of audience.

When writers try to present scientific information in words that the average reader (or layman) can understand, they are popularizing. They use a more informal style than is common in most scientific writing. They try to avoid words with special or technical meanings. When they can't avoid such words, they define them. The article written for laymen (the popularized article), then, is one that anybody can understand without much effort.

Attitudes Toward Popularization

Attitudes toward popularization vary greatly. Some hold that it is a waste of time: some good professionals are too busy to bother about the ignorance of laymen. Furthermore, they say, professionals' reputations depend on other professionals' opinions of their work, not on what ordinary people think. And finally, they assert, professionals cannot write exactly and completely about their subjects without being technical, and therefore they had better not try.

Others hold that such an attitude is mere snobbishness, born of a desire to be superior to the mob. They believe that the really significant discoveries of science—the ones that mark steps in man's progress—can be explained clearly in everyday language. And they insist that such popularization is important—that the healthy state of our culture depends upon popular understanding of scientific knowledge.

A few great scientists have recognized the need for popularization and have gone to the trouble of learning how to popularize. Thomas Henry Huxley deliberately subordinated an ambition to be a great biologist to the task of bringing about a general understanding of the theory of evolution. He was convinced that cultural progress depends upon the increase of natural knowledge and upon the general use of the scientific method by common people.

Types of Popularization

Some of the more common types of popularization are (1) the statement of results; (2) the story of a scientist's work; (3) the story of an observer of science; (4) the direct exposition; (5) the fictitious story with scientific background.

The Statement of Results

The greatest scientific discoveries have been generalizations based upon detailed observations, experiments, or mathematical reasoning. These generalizations can usually be stated in popular terms, even though the bases for them may be highly technical. Thus the Copernican theory, the law of gravitation, the theory of evolution, even the theory of relativity can be explained to the layman, though the techniques employed in establishing them are intelligible only to specialists.

The Story of a Scientist's Work

Telling the story of a scientific achievement is one of the commonest and most successful devices of the popularizer. Any reader can follow with interest and profit the narrative of Pasteur's observations and experiments leading to the discovery of the method of inoculation to control disease, or of Faraday's demonstrations of the properties of electricity, or of the discovery of DNA. If well told, the story will go far toward explaining the meaning of the scientific development.

Story of an Observer of Science

Any writer who is also a keen observer can watch an experiment, a process, a machine, etc., and describe it so that the reader will understand. Any writer, too, can serve as a guide while the reader in imagination takes a trip through a power plant, looks through the electron microscope, or watches a controlled experiment in a psychologist's laboratory.

The Direct Exposition

If the subject is not too technical, the author can explain it by direct exposition. Thomas Henry Huxley thus analyzed the structure of a lobster, slowly building up the reader's understanding of technical terms. Similarly, Alfred Russel Wallace explained how dust in the atmosphere makes our earth inhabitable.

Direct exposition is the method usually employed by writers of textbooks, encyclopedias, and dictionaries. They assume that readers want the information presented and don't need to be entertained while they get it.

The Fictitious Story with Scientific Background

Writers with imagination may inform the reader by means of story situations with scientific background. An astronomer may have thrilling experiences in South America trying to get pictures of a full eclipse of the sun. A Russian spy may attempt to get information about the atomic bomb. A doctor may be faced with two or more alternatives in trying to control an epidemic. If the story is to have value beyond mere entertainment—as it does not in much so-called "science" fiction—it must present exact scientific knowledge.

Techniques of Adaptation for Laymen

In an article entitled "Popularizing Science," from *Science,* M. W. Thistle, chief of the office of the National Research Council of Canada, identifies five barriers obstructing communication between the scientist and the nonscientist:

1. The difficulty of finding symbols to represent technical concepts.
2. The inadequacies of the popular English language for communicating the precise meanings of science, many of which can be expressed only in mathematical symbols.
3. The restrictions of military and political security.
4. The difficulties of getting scientific knowledge into print.
5. The inability of the reader to understand more than half of what he reads.

"When we calculate what gets through the five barriers," Mr. Thistle adds, "it turns out that this is of the order of one ten-thousandth part of what the scientists know."

What devices writers use to catch and hold the attention of laymen readers will depend largely on the subjects they choose.

Choosing an Attractive Title

A good title may arouse the reader's curiosity and stimulate the imagination. If it gives information at the same time, it will have double value. In scientific literature, titles are intended primarily to give information, and consequently often may be long. In the field of popular science, however, titles are intended to catch the attention of perhaps indifferent readers, and are usually short. Typical examples of good titles of books are *Mathematics for the Millions, Microbe Hunters, Snapshots of Science, Creative Chemistry;* of short articles, "Molten Steel," "On a Piece of Chalk," "The Jungle Sluggard," "Leviathan," "The Size of Living Things."

Getting the Readers' Attention

The beginning should be interesting enough that readers are carried on in spite of themselves. Striking statements of fact, incident, dialogue, references to persons—any of the devices of the skilled storyteller may be effective. Here are some typical examples of good beginnings:

> If a well were sunk in the midst of the city of Norwich, the diggers would soon find themselves at work in that white substance almost too soft to be called rock, with which we are all familiar as chalk.—Thomas Henry Huxley, "On a Piece of Chalk."

> Back in 1918 I heard a little girl, nine years old, ask her father this question: "What would the world be like without a war?"—William Atherton Du Puy, "The Insects Are Winning."

> Anyone who joins me in this adventure does so at his peril.—Henshaw Ward, "A Drop of Water."

> "That is the new engineer apprentice for the foundry department," I heard someone say behind the thin wall in the employment office.—C. J. Freund, "Molten Steel."

> In the soft yellow glow of his safety lamp, Campbell peered into the blackness of the tunnel.—Joseph Husband, "Fire-Damp."

> In a prime example of serendipity, one of the principal astronomical discoveries of modern times was made last March by scientists who were not looking for what they found and were, in fact, using a research aircraft designed for an entirely different purpose.—Stephen P. Maran, "Rings Around Uranus."

Personalizing the Language

Scientific writing is likely to be impersonal and objective, but the most readable popular treatments are likely to be personal. The writer emphasizes the "person doing" rather than the "thing done." Names of people who have made scientific discoveries or invented new processes occur frequently; their acts are dramatized. Rudolph Flesch, the author of *The Art of Plain Talk,* made the number of personal references one of the three factors in a formula for determining level of "readability."

Dramatizing the Situation

A story has much wider appeal for the average reader than straight exposition. Whenever writers can communicate their facts through the medium of a story, they are likely to reach more readers. If they can introduce suspense by narrating a series of fruitless experiments, with their accompanying discouragements and frustrations, finally culminating in a great discovery or invention, they will have little trouble in holding readers' attention. If they can explain the work of a foundry by narrating the experiences of an apprentice engineer on the job—as C. J. Freund did in "Molten Steel"—they will have communicated information almost without the readers' being aware of it.

Such indirect methods are of course less economical than direct exposition. The author of a textbook in physics who tried to present all the facts of physics in story form would probably need three times as much space as in direct exposition. For students who *want* to learn those facts, presenting them in dramatic form is unnecessary.

Appealing to the Senses

Scientific description is usually limited to such objective qualities as size, shape, materials, finish, and connections. Little attempt is made to appeal to readers' imaginations. The popularizer, on the other hand, tries, as Joseph Conrad, for example, tries, to make us see every detail of the last hours of the old sailing vessel whose cargo of coal has caught fire in the Indian Ocean.

> I heard all round me like a pent-up breath released—as if a thousand giants simultaneously had said Phoo! and felt a dull concussion which made my ribs ache suddenly.—"Youth."

Like a painter, the writer splashes narrative with color:

> The sea-anemones first attract attention, showing as splashes of scarlet and salmon among the olive-green seaweed, or in hundreds covering the entire bottom of a pool with a delicately hued mist of waving tentacles.—William Beebe, "Secrets of the Ocean."

The writer tries to give an impression of the unfamiliar by comparing it with the familiar:

> The crippled ladle was carried away from the ingot mold, and the stream of metal splattered to the floor and bounced back again, not like water but more eagerly, like thousands of small incandescent rubber balls.—C. J. Freund, "Molten Steel."

Analogy may serve the same purpose, as when Simon Newcomb tries to give us a conception of distances in the universe by comparing the universe with a model:

> Let us imagine that, in this model of the universe, the earth on which we dwell is represented by a grain of mustard seed. The moon will then be a particle about one fourth the diameter of the grain, placed at a distance of an inch from the earth. The sun will be represented by a large apple, placed at a distance of forty feet. . . .—*Astronomy for Everybody.*

It is hardly necessary to add that such description is more difficult to write than the objective and factual, like the following:

> The blade of the T-square is made of a long strip of maple with celluloid edges through which the draftsman can see his work.

Explaining Technical Terms

As far as possible, anyone writing for laymen should avoid highly technical terms. Scientists writing for others in their own field may assume that readers will understand technical terms; but popularizers must assume that readers will not understand them. When there is no substitute, then careful popularizers define such words when they first use them.

For bright light sources—such as the Sun, laboratory flames, arcs, sparks, and glows—it has been found desirable to substitute for the prism a diffraction grating. This is a flat plane of hard metal which has been minutely furrowed with a series of delicate parallel lines cut into the surface, sometimes as many as 30,000 to the inch.—George W. Gray, "Eyes."

At the beginning of a discourse particularly, authors avoid using technical terms. Not until their readers are well oriented will they begin using them, and then one at a time, explaining carefully as they proceed.

* * * * *

All of the techniques so far discussed are variants of the general principle *Remember the reader.* So long as the writer keeps the reader clearly in mind and makes things as easy as possible to understand, communication is likely to be successful.

EXERCISES

A. *Judge the effectiveness of each of the following beginnings and titles of popular-science articles*:

1. Today, as never before, the sky is menacing.—Loren C. Eiseley, "Little Men and Flying Saucers," from *The Immense Journey.*
2. Somebody once went to a lot of trouble to find out if cigarette smokers make lower college grades than nonsmokers. It turned out that they did.—Darrell Huff and Irving Geis, "Post Hoc Rides Again," from *How to Lie with Statistics.*
3. In August 1946 a dry lightning storm, a thunderstorm without rain, started a fire in a remote part of the Boise National Forest in Idaho.—Bernard DeVoto, "The Smokejumpers," from *The Easy Chair.*
4. Referring to wildlife management, Aldo Leopold, acknowledged "father" of the science, once said, "Perhaps the day will come when many states will learn that research cannot be turned on and off like a spigot."—Durward L. Allen, "Of Skunks, Facts, and Time," from *Discovery: Great Moments in the Lives of Outstanding Naturalists.*
5. From the rather self-conscious heights of our own state of equivocal civilization and of that of the community to which we belong, we men of the latest period of human development have traditionally taken the view that whatever has preceded us was by so much the less advanced.—Ashley Montagu, "The Fallacy of the Primitive," from the *Journal of the American Medical Association.*
6. Automobiles, like women, can be fun, but they are expensive.—John Keats, "Ask the Man Who Owns One," *Atlantic Monthly.*
7. The most advanced branches of science, namely, the physical, have developed their own special languages, such as the mathematics of theoretical physics and the molecular structure symbolism of chemistry.—Anatol Rapoport, "The Language of Science: Its Simplicity, Beauty, and Humor."

8. The most remarkable discovery made by science is science itself.—J. Bronowski, "The Creative Process," from *Scientific American.*
9. In my student days I kept hearing tales of one of the professors—I shall call him Z, as he is still going strong, and I have no wish to embarrass him—who was a wonder for the amount of high-grade scientific research which he turned out apparently without the slightest effort.—Eric Temple Bell, "The Search for Truth."
10. A man scaled down to the size of an atom would find a trip through a crystal a tedious expedition.—B. D. Cullity, "Aligned Crystals in Metals," from *Scientific American.*
11. Among those who passed through the general clinic of Lenox Hill Hospital, at Seventy-sixth Street and Park Avenue, on Monday morning, April 6, 1942, was a forty-year-old Yorkshire dishwasher whom I shall call Herman Sauer.—Berton Roueché, "A Pig from Jersey," from *Eleven Blue Men.*
12. In the long and lurid years since the first experimental scientist made the contents of a beaker bubble and hiss, no concoction man has brewed has so rapidly, so thoroughly, or so theatrically permeated the world as a white, crystalline, odorless, tasteless nerve poison that is known in scientific literature as the para-para-prime isomer of dichloro-diphenyl-trichloroethane, a chlorinated hydrocarbon—and everywhere else as DDT.—Robert Rice, "DDT."
13. We human beings are only fairly good at killing.—Gove Hambidge, "The New Insect-Killers."
14. Sloths have no right to be living on the earth today; they would be fitting inhabitants of Mars, where a year is over six hundred days long.—William Beebe, "The Jungle Sluggard."
15. In an article published a few years ago, the writer intimated with befitting subtlety that since most concepts of science are relatively simple (once you understand them) any ambitious scientist must, in self-protection, prevent his colleagues from discovering that his ideas are simple too.—Nicholas Vanserg, "Mathmanship."
16. Two hundred and fifty years ago an obscure man named Leeuwenhoek looked for the first time into a mysterious new world peopled with a thousand different kinds of tiny beings, many of them more important to mankind than any continent or archipelago.—Paul de Kruif, "Leeuwenhoek."

B. *Make a combination outline listing the four main types of audience discussed above as the major topics. Then under each type, outline the other qualities of each audience. Be as complete as you can. Make another outline which corresponds with that one point-by-point, this time listing what an author can do to meet that particular audience characteristic halfway.*

Notes

1. See Thomas E. Pearsall, *Audience Analysis for Technical Writing* (Beverly Hills, California: Glencoe Press, 1969) for Pearsall's complete discussion of this.

4

Remember the Reader II: Headings and Visuals

Despite the limitless variety of kinds of audiences, some important generalizations may be made about adaptation techniques for all types of readers. One such generalization is the audience's need for writing with clear structures (discussed in chapter 2). That need, in turn, produces the writer's need for planning (also discussed in chapter 2). Other ways to reduce the amount of effort which reading your writing demands are discussed below. The abundant use of headings and visuals distinguishes scientific writing from most kinds of popular writing, giving it qualities of orderliness and visual impact all readers, of whatever kind, can appreciate.

STATEMENT OF THE CENTRAL IDEA

The old rhetorics—and some of the new ones—list the basic requirements of good exposition as Unity, Coherence, and Emphasis, and of these the most important is unity. To achieve it, you must succeed in ordering all of the elements of your composition so that each contributes to a unified effect. The principle is still of first importance.

The *central idea* of the whole must be made clear to your reader. Usually you should state it, if not in the first sentence, at least in the first paragraph. Only in special cases may it be desirable to defer statement of the central idea to the middle or the end. (See Deductive and Inductive Arrangements in the chapter on Definition.) The introduction to a long report begins with a statement of the object of the report, or of the investigation, study, experiment, or observations described. After stating the central idea or purpose, you can also help your reader if you tell what your plan is. If there is a table of contents, as in the long report, it should give the kind of preview needed.

THE VALUE OF HEADINGS

Use a topical heading at the beginning of each division of the paper.[1] Such headings are common in technical writing and are becoming common in all scientific writing. They have several values:

1. They enable readers to find easily any part they want to read. With them, readers will waste no time hunting. Textbooks, scientific papers, and especially reports are often used for reference, and are not necessarily read through from beginning to end. The absence of headings would lead to irritation and loss of time while readers grope around and try to find what they want. Compare the problem of students trying to find what they want to review in a textbook.

2. They make less necessary the use of transitional words, phrases, sentences, and paragraphs. A topical heading is the equivalent of a signpost indicating the

end of one section and the beginning of another whose subject it announces. Though transitional devices may still be helpful at times, they do not catch the eye as do headings.

3. They enable the reader to tell, from a glance over the paper as a whole, what the plan of organization is. The headings constitute the items of a topical outline.

4. The use of headings almost forces you to arrange your material logically. When you use them, you are less likely to put your facts on paper in the haphazard order in which they first occur to you. The actual writing becomes much easier when you plan the sequence of headings and subheadings from the beginning.

Suggestions for Proper Use

Several bits of advice that may make your use of headings more profitable are the following:

1. Between the title and first heading, or between a main heading and any following subheadings, write at least one transitional sentence, perhaps a short paragraph, preparing your reader for what follows: a statement of central idea, a definition, the plan of division to follow. The sentence preceding this paragraph is an example.

2. In the sentence following a heading, do not use a pronoun whose antecedent is in the heading. The headings are *outside* the text; they are signposts, as it were, and not grammatical parts of your exposition. For example, if the heading is *Jet Engines,* the following sentence should be, not "These are becoming increasingly efficient," but "Jet engines are becoming increasing efficient."

3. Keep headings that are logically equal and parallel in the same grammatical form. For example, if the main heading is *Analysis of the Building Site* and subheading 1 is *Topography,* subheading 2 should be, not *Testing the Water Supply,* but *Water Supply.* The grammatical form of the first heading in a series is the pattern for the whole series. Choose the form in which all the topics in the series can be most easily expressed. The grammatical form may vary from one series to another. Main headings may be nouns, and subheads some other construction. This rule is merely an application of the principle of parallelism (see chapter 20) to the use of headings.

4. Limit the length of text matter under a heading, especially in reports. If you find any section extending more than a page or so in double-spaced typed form, see if you cannot break it down into subsections with subheads. Your reader shouldn't have to look through several pages of text searching for the fact wanted.

Form and Layout

With the typewriter you can indicate the relative importance of headings in the four different ways listed below.

1. By spacing and capitalizing the letters (M A I N H E A D I N G)
2. By using capital letters only (SUBDIVISION)
3. By underlining (Minor Subdivision)
4. By putting headings in different positions (centering, placing at the margin, indenting, etc.)

In general, the more important the heading, the more of these ways of indicating relative importance you should use. Notice the application of this principle in the following system, which provides for six levels of importance:

M A I N H E A D I N G	(spacing, capitalizing, underlining, centering)
MAJOR SUBDIVISION	(capitalizing, underlining, centering)
MINOR SUBDIVISION	(capitalizing, centering)
Section of Minor Subdivision	(underlining, centering)
Subsection	(underlining, placing at the margin)
Paragraph Heading. – Paragraph beginning	(underlining, indenting)

So complicated a system will rarely be needed.

When you are distinguishing fewer degrees of importance in headings, you can devise simpler systems like the following in which four levels are provided:

MAIN HEADING

MAJOR SUBDIVISION

Minor Subdivision

Paragraph Heading. – Paragraph beginning

Where you have a single series of headings at the same level, you can set them in whatever way you prefer—center heads, side heads, paragraph heads, etc.

In longhand, the problem of distinguishing headings and making them stand out prominently on the page is more difficult. Printers have no problem because they have many different sizes and styles of print available.

However complicated or simple your system of headings may be, remember that they should stand out on the pages so that they instantly catch the reader's eye. They are devices primarily to help readers find what they want and to indicate the plan of organization.

Numbering and Lettering

The use of letters and numbers with headings (see chapter 6 on Classification) is optional. If you need only a few headings, the symbols are of doubtful value; if you need many, they may be helpful to your reader. The symbols most commonly used at the different levels are these: first level, Roman numerals; second level, capital letters; third level, Arabic numerals; fourth level, lowercase letters; fifth level, lowercase letters with accent (or prime) marks—thus: I, II, etc., A, B, etc., 1, 2, etc., a, b, etc., a′, b′, etc.

Another system of symbols, sometimes called the decimal system, consists of Arabic numerals for all levels, with periods separating them. In this system a symbol like 3.1.6 before a heading or without a heading would mean "third main division, first subdivision, sixth section." The system facilitates reference in complicated analyses. (For an example of its use see the section on Punctuation in chapter 19 on Mechanical Aspects of Style.)

THE VALUE OF VISUALS

One of the best ways to make any scientific paper or report clearer, more attractive, and easier to read is to make free and frequent use of visuals. Your use of a table, graph, or picture can help your reader in at least these three ways: The visual can emphasize one particular item in a mass of data, such as when a table is printed with the key numbers in bold-face type. Your visual can also be used to provide additional detail beyond that contained in the text, such as when a graph charts figures for a longer period of time than the text iself covers. Finally, visuals can make your text more attractive and appealing to the reader, demonstrating clearly that your writing takes the reader's needs and interests into consideration.

Types of Visuals

There are basically three methods of presenting information visually to supplement statements in sentences and paragraphs.

1. *Tabular methods*: sets of related numbers, symbols, or items of any sort arranged for ease of reference or comparison, often in parallel columns. *Example*: the chart comparing the method of science with the process of writing (Table 1 in chapter 2).
2. *Graphic methods*: diagrams, figures, curves, and graphs based upon collected data. *Example*: the flow-chart representing the recursiveness of the writing process (Figure 1 in chapter 2).
3. *Pictorial methods*: maps, sketches, drawings, and photographs representing the actual appearance of areas and objects, or the envisioned appearance of abstract concepts. *Example*: the two-stage representation of the definition process (Figures 1.1 and 1.2 in chapter 1).

These methods supplement each other as well as the text. It is easier to compare numerical data when they are presented in graphical form than when they are arranged in tables. A photograph may give the reader a better understanding of an object or an area than a mechanical drawing or a map. Language is an inadequate medium for presenting many kinds of information; at its very clearest it often fails to do what a graph or a picture can do easily.

On the other hand, tables, graphs, and drawings nearly always need some verbal explanation. A figure that seems crystal clear to you because you are familiar with what it represents may be a puzzle to the reader. The significance of even the simplest picture should be made clear in the text of the report, on the same page if possible, otherwise on the preceding or the following page.

Almost 50 examples of visual aids occur in this book. The names of those listed will be found under the heading "Visual aids" in the index. Some not illustrated, but frequently found in books, newspapers, and magazines, are: photographs, bar diagrams, pie diagrams, pictographs, and exploded drawings, often printed in two or more colors.

For more good examples of such methods of presentation see almost any issue of *U.S. News and World Report, Psychology Today, Scientific American,* or annual reports of business corporations. For fuller treatment of the use of graphic aids, see any of the books given in the bibliography.

Suggestions for Proper Use

In the preparation of illustrations—whether tabular, graphic, or pictorial—you should conform to certain conventions.

1. Number illustrations consecutively and give them adequate titles or "captions"; for example, "Figure 1. Completed Sand Mold." Even if there is only one illustration, give it a number and title.
2. Write the number and title *below* the illustration.
3. Keep the reading matter on illustrations to a minimum. Spell out all words in full where space permits. If you use abbreviations, use those recommended (especially in technical writing) in "American Standard Abbreviations for Scientific and Engineering Terms" (ASA Standards Specification N.Z10i). The list is reprinted in chapter 19.
4. Use reference letters or figures on any drawings if they will make it easier to correlate the description with the drawing.
5. Put each drawing as close as possible to the accompanying description.

EXERCISES

A. *You can get good practice in adapting your language to readers' needs by explaining a different subject for several types of reader. Philip W. Swain, once editor of* Power, *in an article "Giving Power to Words" in the* Journal of American Physics, *October 1945, suggested such an exercise for physicists:*

First write a sedate and technical little treatise on the gyroscope—say, 1000 words. You might entitle it "A concise summary of the physical principles underlying gyroscopic phenomena." Then rewrite (and retitle) the piece seven times for the following seven types of audience or reader:
The Latin faculty of your university
Some imaginary university president to be impressed by your profundity
A student of first-year physics
The *Reader's Digest*
A Rotary Club meeting
A mechanic skilled but unschooled
An eager 10-year-old boy, interested in gadgets

Clearly there would be considerable differences in the seven write-ups. Try the exercise yourself, making sure that you adjust your use of headings and visuals for each different audience.

B. *Write a paper on the uses, forms, and effectiveness of headings in one of your textbooks.*

C. *Compare the uses, forms, and effectiveness of the headings in two or more of your textbooks, and report your findings.*

Notes

1. An introductory paragraph at the beginning of a paper or long section of a paper (unless it is in long-report form) needs no heading. Your reader will expect the paragraph to state the central idea and perhaps tell him how you plan to proceed. Such a heading as "Introduction" or "Discussion" is almost meaningless.

Scientific Papers

5

Definitions

Definition is frequently essential to effective communication, but nowhere more so than in scientific papers and reports, where "to tell the truth, the whole truth, and nothing but the truth" is the first requirement. Thoreau remarked that "it takes two to speak the truth, one to speak and another to hear." The obligation of speakers (or writers) is the heavier of the two. They cannot shirk that obligation if they hope to be understood.

THE NATURE OF LANGUAGE

A language is a system of symbols which people by common agreement use to represent things and communicate thoughts. Most of these symbols (words) convey a specific meaning to the persons familiar with them. The "meaning" of a word is the particular object, idea, or action that it calls up in the mind of the person who hears it (or who sees the marks that represent it on paper). "Stop," "ouch," "horse," "baby," "black"—such symbols are rarely misunderstood. Their meanings are so generally agreed upon that communication with them is fairly sure, though never completely so. The meaning of "horse" for you may be "racehorse crossing the finish line" while for another it is "Dobbin rolling in the barnlot."

Other symbols used almost as frequently have meanings far less certain. "Love," "worship," "God," "democracy," "cute," "conservative"—any of these words is quite likely to mean one thing to you but something altogether different to your roommate. "To worship God' may signify to you going to the Methodist church, singing hymns, and listening to a sermon; to another it may signify quiet solitude and reflection on a mountain top. An argument between you and your roommate about worship might result in a violent quarrel, which a little previous discussion of meanings might have avoided. To you "democracy" may mean complete freedom to do as you please; to your roommate it may mean a society in which the individual voluntarily gives up most of his "rights" for the good of the whole. People fighting "for democracy" may quite possibly be fighting for a thousand different things that the word "means" for them.

Technical words, on the other hand, are less likely to be troublesome, because the people who use them are careful to define their meanings. Although "gyrose," "meiosis," and "codeclination" may have no meaning at all for the average person, they signify something quite specific to the scientists who use them. Popularization has blurred the meaning of some scientific terms. Imagine the difference in the meaning of a word like "relativity" for Albert Einstein and for the average person on the street. "Evolution" in popular usage has become so perverted that it is still a "fighting word" for thousands of people. "Communism" has undergone a similar change.

THE IMPORTANCE OF DEFINITION

The main purpose of language, especially in scientific writing, is to communicate. If the symbol (word) that you use means one thing to you and something quite different to your reader, communication fails. If you write, "The actor in Greek tragedy wore cotherni," the chances are good that your reader won't know whether a "cothernus" is a hat or a cloak or a sword or the high laced boot that it happens to have been. If you say, "I believe in God," and the symbol "God" means to you a benevolent white-bearded old man of heroic proportions sitting on a big gold chair somewhere above the clouds, and to your listener an impersonal force or energy that permeates all matter, you have not communicated. If you say, "Bill got hit by a hammer," and you mean by "hammer" the weight used in the hammer-throwing contest of a track meet, and not a carpenter's hammer, you will have communicated only a fraction of your thought.

It is necessary, therefore, if you would speak or write with any certainty of being understood, to make sure that the symbols you use mean the same thing to your reader or listener that they mean to you. To attain such precision requires constant definition.

In technical language especially, the need for definition is likely to be frequent. It is true that the meaning of most technical terms is clear to those accustomed to them. All technicians who use such terms as "oxygen," "rhombus," "volt," "hydrometer," and "louver boards" will define them in the same way. But others may get the wrong meaning or little meaning from them. One of the most common failings of technicians is to assume that others understand the technical terms they use.

The more extensive and specialized your knowledge becomes, the more danger there is that your words will not be generally understood. Finally the point is reached where the terminology of one specialist is understood only by another specialist in the same field—sometimes not even by him. As T. V. Smith, former professor of philosophy at the University of Chicago, once wrote:

> When I go to the American Philosophical meetings I don't find it worth my time to listen to half the papers that my colleagues read because I do not, frankly, know what the guys are talking about and, being human, I half suspect that they don't. . . . Without meaning to do so, the world of science, of the professors, glories in the invention of technical languages that connote the cutting up of the solidarity of mankind until it gets very, very difficult to follow the general argument of science itself.[1]

There are two ways to solve this problem. One is to avoid highly specialized technical terms and try to express meanings with the symbols familiar to most readers; the other is to define the technical terms used. The first solution is often undesirable because the familiar symbols—the nontechnical words—are likely through long use to have become worn and inexact. They have many "referents," that is, things, ideas, etc. which they may symbolize. The second solution is preferable, even though it may be unnecessary for technically informed readers. They can ignore a list of definitions or interpolated definitions that other readers find essential to their understanding. In either solution the need for definition is imperative.

THE NATURE OF DEFINITION

The word "define" comes from the Latin *definire*, "to mark the limits or boundaries of." Defining a word, then, consists simply of "fencing it in," telling what referents it symbolizes. Where the referent is concrete, something in the physical world, the most certain method of definition is to point out the referent: "black" is the color of the print on this page; a "book" is the bound sheaf of printed sheets of paper from which you are reading. So long as we can use our senses to identify the object symbolized, uncertainty of meaning is unlikely.

Words range all the way from those with only a single referent (Lake Erie) to those with no referent identifiable by the senses (nobility, force); and the higher up the abstraction ladder[2] the less probability there is of identifying the referent and the greater need for verbal definition. If you are trying to tell someone what a "sextant" is, and no sextant is available to point to, then you must resort to words: "an instrument for measuring altitudes of celestial bodies from a moving ship or airplane with a maximum angle of 60 degrees between its reflecting mirrors."[3] If you are trying to define "house," a class name for which a great variety of referents may be listed, it is not enough to describe one particular house. Through a process of abstraction and verbalization you must determine the characteristics common to all houses: "a structure intended or used for human habitation: a building that serves as one's residence or domicile esp. as contrasted with a place of business: a building containing living quarters for one or a few families. . . ."[4] If you are trying to define "glory" or "progress," words for which there are no concrete referents, then your only resource is to use words. You cannot point to or picture anything that "is" glory or progress.

A word stands in the same relation to its referent as a map stands in relation to the territory it represents. To be of value to the traveller, a map must represent clearly the region which it "symbolizes"; in the same way a word, if it is to aid in communication, must represent accurately the thing it symbolizes.

THE STEPS IN DEFINITION

When it becomes desirable to "mark the limits or boundaries" of a word, logical procedure, first described by Aristotle in the fourth century B.C., requires two steps:[5]

1. Assign the thing being defined to a class (*genus*), a group of things having similarities.

2. Enumerate the special qualities of the thing being defined which differentiate it from other members of the class (*differences*).

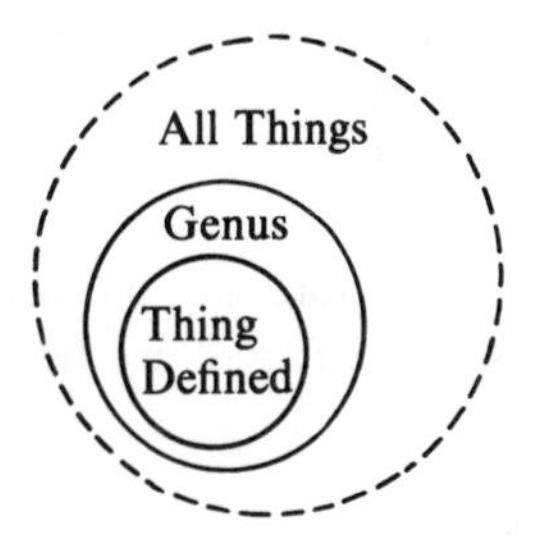

Figure 5.1. Areas in definition.

The process may be represented graphically as in figure 5.1: The outer dotted circle is an indeterminate area including all possible referents. Assigning a thing to its *genus* excludes most of this area; listing the *differences* puts up a fence which separates it from others of the same genus.

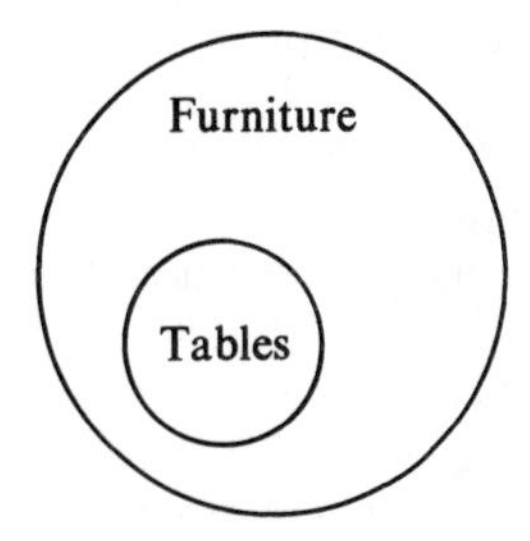

Figure 5.2. Genus and differences.

For example, in defining "table" you can first put it in the genus "furniture" (figure 5.2) and then note that it differs from other articles of furniture in construction and function: "A table is an article of furniture having a smooth flat top fixed on legs and used to set objects on."

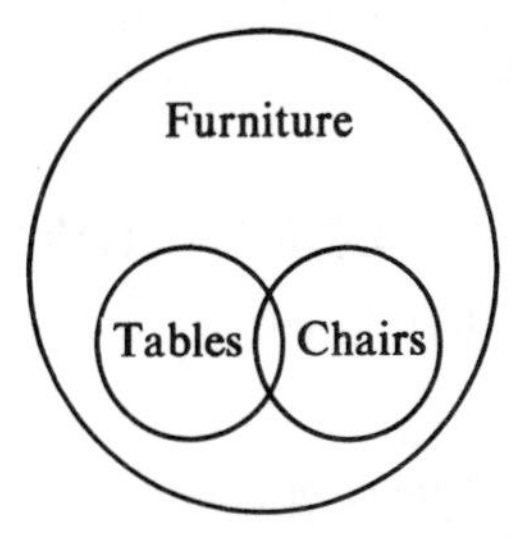

Figure 5.3. Faulty statement of differences.

A "chair" you could define as "an article of furniture with a single seat and a back." If, however, in defining "table" you had noted merely the one difference "fixed on legs," your definition would have been incomplete because chairs and stools and cabinets also have legs. The two small circles in figure 5.3 would then overlap.

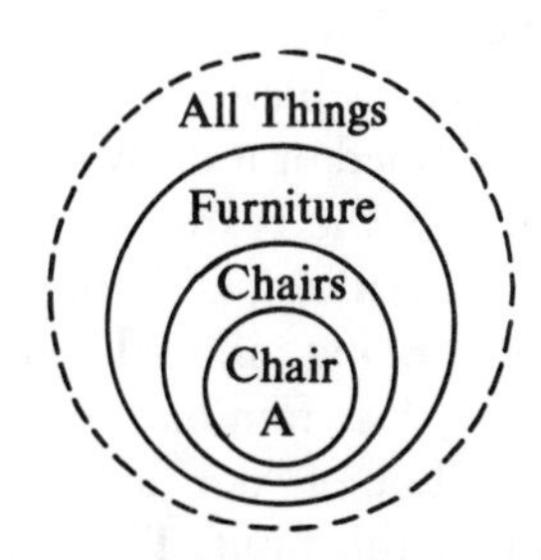

Figure 5.4. The narrowing process.

You can carry the process of definition a step further by adding individual differences more and more specific until the definition applies to only one particular member of the class; for example, "an article of furniture with a single seat and back, the parts being made of hickory and assembled by my grandfather in 1842." Figure 5.4 represents the narrowing process.

The procedure is illustrated in the parlor game of "Twenty Questions." Usually the subject to be identified in twenty questions is assigned to one of three categories: animal, vegetable, or mineral. Skillful questioners try to narrow this category to a *genus*: from "animal," for example, to "American, dead." Then they try to find the *differences*—sex, profession, place of residence, etc.—and finally identify the subject as Daniel Boone.

To illustrate further the two steps in logical definition, a few examples from *The American Heritage Dictionary of the English Language** are listed:

Word	Genus	Differences
bertha	a collar	A wide, deep collar, often of lace, that covers the shoulders of a low-necked dress.
cantilever	a beam	A beam or other member projecting beyond a fulcrum and supported by a balancing member or a downward force behind the fulcrum.
dachshund	a dog	A small dog of a breed native to Germany, having a long body with a usually short-haired brown or black and brown coat, drooping ears, and very short legs.
fasces	an emblem	A bundle of rods bound together about an axe with the blade projecting, carried before magistrates of ancient Rome as an emblem of authority.
stalagmite	a deposit	A cylindrical or conical deposit, usually of calcite or aragonite, projecting upward from the floor of a cavern as a result of the dripping of mineral-rich water.

THE WEAKNESSES OF LOGICAL DEFINITIONS

A few weaknesses of logical (Aristotelian) definitions are so common that they deserve special mention:

1. The definition is in words and therefore at least one step up the abstraction ladder from the thing itself. As Anatole Rapoport explains[6], a good Aristotelian definition can be framed for a fiction, like "vampire": a vampire is a person who habitually sucks other people's blood." The "differences" listed in the definition should be as concrete and specific as possible.

2. The definition may not include the word defined or a derivative of it. To say that "a jet engine is an engine" is useless unless the reader already knows what an engine is. To say that "criticism is the art of criticizing" is almost a waste of words, as it is to say that "inflation is the state of being inflated."

3. The definition may not be stated in words less familiar than the term being defined. Dr. Samuel Johnson's famous definition of "net" as "a reticulated structure with small interstices" is an illustration of this fault. The definition of "argillite" in the Second Edition of *Webster's New International Dictionary,*[7] "an

argillaceous rock, differing from shale in being metamorphosed and in lacking fissility," can profitably be compared with that in the Third Edition,[8] "a compact argillaceous rock differing from shale in being cemented by silica and from slate in having no slaty cleavage."

4. The definition may not consist merely of a synonym. If it does, it is not a complete Aristotelian definition. Some smaller dictionaries have this fault. Though synonyms may be helpful, they are usually unsatisfactory because no two words mean exactly the same thing. "Impecunious" means something like "poor," and "lawful" means something like "legal," but the finer shades of their different meanings need to be stated.

5. The definition may not be allowed to evaporate into mere talk and vague abstractions. To define "courage" as "when one encounters dangers without fear" and "liberty" as "freedom from external or foreign rule" is almost a complete waste of words. To say that an abstract painting is one in which "there is nothing similar to anything in the real world" is negative talk, telling us nothing about what abstract painting *is*.

6. A definition beginning "something is when" is inadequate because no *genus* is named. The "when" clause is almost invariably a statement concerning a *difference*. "An accident is when something happens that nobody intended" illustrates this fault: the genus, an *event*, is omitted.

THE OPERATIONAL DEFINITION

For terms designating things or processes or procedures it is sometimes desirable to substitute for the logical definition what has been called an *operational* definition, or a *definition by prediction*.

Such a definition says in effect, "if you will do so and so, you will find out what something is." It is at the lowest level of abstraction. For example, a short circuit might be defined without words merely by demonstrating what happens when a side circuit of low relative resistance is made to connect two points in a circuit of higher resistance. The meaning of "cavity" could be communicated without words, merely by pointing to a hole in the ground. In words, the nearest approach to the operational definition is a direction telling the reader what to do in order to observe the thing being defined.

Of the methods of expanding logical definitions explained in the following section, the specific example or illustration is similar to the operational definition.

EXPANDED DEFINITIONS: TEN METHODS

Of the five definitions quoted from *The American Heritage Dictionary* earlier in this chapter, probably none would be completely satisfactory to the average reader, who would welcome more information than the logical definition gives. Who was the *Bertha* after whom the collar was apparently named?

What are *cantilevers* used for? Is a *dachshund* good for anything? How is the word *fasces* related to *Fascist*? What are *calcite* and *aragonite*?

There are several methods by which you can expand the logical definition. Sometimes you may need to use only one of these methods, sometimes a combination of them.

Explication

In using the method of explication you explain the meaning of key words in the logical definition. For example, you might expand the logical definition of "engineering"—"the application of science and mathematics by which the properties of matter and the sources of energy in nature are made useful to man"[9]—by explaining the meaning of "science," "mathematics," "properties," "matter," and "energy."

This method has only a limited application. When you use it, you ordinarily should combine it with other methods.

Analysis, or Division

Analysis refers usually to the separation of a whole into its component parts: division refers to the grouping of the particulars that make up a class. For example, a definition of "air" as a mixture of certain gases that form the earth's atmosphere might be followed by an analysis of the mixture, listing and defining each of the gases that are a part of it. A definition of "mammal" as a class of higher vertebrates that nourish their young with the milk of females might be followed by a listing and defining of the divisions of the class. The method requires application of the principles of classification stated in chapter 6 of this book.

Since the definition of almost any noun can be expanded by analysis or division, the method is widely used. For detailed explanation of the principles of analysis and examples of their application see chapter 7 of this book.

Description (Giving Details)

A common method of expanding a logical definition is to describe the thing being defined. To give the details of mechanisms, processes, organizations, and objects is likely to help the reader to a better understanding of them. Basically the method is that of analysis, subordinated in an expanded definition to the explanation of what the thing is. For example, a definition of a recorder or a bagpipe might well be supplemented by a description of it.

Illustration

Illustrating the term being defined by an example is one of the most helpful and interesting methods, though it may require several illustrations to give an adequate conception of a general or abstract term. In a definition of "bravery" an example of valiant conduct on the battlefield would be inadequate and misleading, because bravery can be manifested in many other situations: by a mother protecting her child, by a business person defying a racketeer, by a young person opposing the weight of public opinion. You should therefore consider whether the illustrations are typical.

Pictures and Diagrams

Though it is not always true that "a picture is better than a thousand words," any user of a good dictionary will agree that a picture or diagram is often helpful. If you are defining something that can be represented by a graphic method, you should use whatever pictorial devices will be most useful and interesting for your reader.

Comparison, Contrast

In using the method of comparison or contrast you define *A* by comparing it with *B*, perhaps *C*, *D*, etc. For example, you can expand a logical definition of "architect" by comparing the architect's work with that of a contractor, a builder, and an engineer. The method is especially helpful when you compare something unfamiliar with something familiar. You can best define "communism," perhaps, by showing how it is like and unlike democracy and dictatorship. Thomas Henry Huxley used the method when he showed how the scientific method is like and unlike our ordinary thinking.

Analogy

Analogy is a special type of comparison in which there is resemblance in one or more particulars between things that are otherwise unlike. Thus death is analogous to sleep in that there is cessation of activity and of consciousness. Waves of sound are analogous to waves of water in that they spread out from a center.

Metaphors and similes also involve comparison, but they are usually short and limited to a single concept ("like living in a bean pod," "well-upholstered ladies"), whereas the analogy may be much longer and more detailed. For example, some economists find an analogy between the debts and the spending of a family and those of a nation, and carry the analogy to the point where questionable inferences are made.

A good analogy may be helpful, especially in explaining some difficult scientific or economic hypothesis. The danger of its use is the tendency to infer that because things are alike in one or more respects, they are alike in all. How valid, for example, is the analogy between the changing of horses while crossing a stream—which an old proverb forbids—and voting for a new leader in a political campaign?

Elimination

By the method of elimination you may help to define a thing by telling what it is *not*. Used alone, it could not provide an adequate definition; but when combined with other methods it can be effective, especially when you need to correct wrong conceptions. For example, you might begin a definition of democracy by eliminating the wrong conceptions before stating affirmatively what democracy is.

Description of Origin, Causes, Effects

Often you can help to define a term clearly by showing how it developed or by describing its causes and effects. An expanded definition of "collective bargaining" probably should include an account of how it began. Some things about which we actually know little (electricity, viruses) can be explained best by reference to their causes and effects.

History of the Word (Etymology)

Explaining the original meaning of roots, prefixes, and suffixes often is illuminating, especially now that Greek, Latin, and Anglo-Saxon, the languages from which most of our English words are derived, are so little studied. "Cybernetics," the comparative study of the control systems of living bodies and of man-made systems, is derived from the Greek word meaning to steer or to govern. "Sabotage," the deliberate slowing down or interruption of a manufacturing process, is derived from "sabot" (wooden shoe) and developed from the French workmen's use of wooden shoes to damage machinery and stop production. Though a word may outgrow its original meaning, as "sabotage" has, the explanation of its etymology is often essential for a complete definition.

DEDUCTIVE AND INDUCTIVE ARRANGEMENTS

In an expanded definition you will usually want to begin with the logical definition and then expand it by using some of the methods listed. This arrangement is called *deductive*, because it begins with a general statement—the logical definition.

An alternative arrangement is to begin with an illustration, a comparison, an analogy, or one of the other methods of expanding, and reserve the logical definition for the end. Since this arrangement begins with particulars and ends with a general statement, it is properly called *inductive.*

A third possible arrangement is to place the logical definition somewhere in the middle, preceded and followed by use of one or more of the methods of expansion. This arrangement combines the deductive and inductive procedures.

PROBLEMS WITH DEFINITIONS

Like any other verbal pattern, the definition pattern has typical variations, as shown above. It also presents typical problems. Anyone who writes an expanded definition needs to consider the following three typical problems.

Definition Too Broad

You must make certain that your definition only includes the objects proper to it. You cannot rely on your reader's knowledge of the subject to exclude things that to you may seem obvious. If I define "shuttle" as a people-mover that travels back-and-forth between two or more points on a regular basis, I have carelessly included elevators (and the bicycle I ride to school every day) in my definition. Once you see you have this kind of problem, you can solve it by returning to the narrowing process.

Definition Too Narrow

At times you may inadvertently exclude part of the class you want your definition to name. If I define "fuel-efficient engines" as those engines which average at least 25 miles per gallon of gas in town and 35 on the highway, I may leave my reader wondering whether and where steam, electric, solar, and all the other possible varieties of engines fit into my scheme. Once again, when you write an expanded definition you must be sensitive to the fact that the reader comes to it with his or her own implicit assumptions and beliefs, which may be very different from your own. Be sure your definition draws its boundaries the same way for the reader that it does for you.

Definition Poorly Written

Implicit in the very nature of definition is the overuse of sentences whose verbs are all forms of "to be." Though it may be impossible to write a definition without using a form of "to be" somewhere, there is no more abstract or hard-to-read page than one with a majority of sentences using "to be" verbs. When you revise your expanded definition, be sure to include one step in which you check your verbs. If three out of every four are forms of "to be," you will want to change them.

EXERCISES

A. *Criticize the following attempts at logical definition:*

1. A kernel is the seed inside the strong endocarp of a drupe.
2. An amateur is a person who is not a professional.
3. A strike is a situation in which employees refuse to work.
4. Canned goods are things sold in cans.
5. A lobbyist is a person who tries to influence the vote of legislators.
6. Courage is the quality of being fearless or brave.
7. A hammer is a tool for driving nails.
8. A forward pass is a play in football in which the ball is thrown to someone beyond the line of scrimmage.
9. Oats are "a grain which in England is fed to horses, but in Scotland supports the people."—Dr. Samuel Johnson.
10. A sonnet is a 14-line poem in iambic pentameter.
11. An octagon is a plane figure with eight sides and eight equal angles.
12. A hat is an article of clothing worn on the head.
13. A baguette is a small moulding that is like but smaller than the astragal.
14. Fear is when one anticipates danger.
15. An occluded front is the air front established when a cold front occludes a warm front.

B. *Write a one-sentence definition of one meaning of each of the following terms:*

altimeter, anger, azimuth, bamboo, calendar, caliper, calorie, dead center, denture, dormer, echelon, eclipse, electromagnet, frustum, parquet, peavy, tap, viaduct, vibrato, willowware

C. *Write an expanded definition (400-500 words) of one of the following terms:*

anachronism, astrology, biophysics, business ethics, conservative, deductive reasoning, digital computer, flood control, freedom of speech, gerrymander, hydrometer, insulator, liberal, modern music, poetry, selling on margin, socialism, solecism, spaceship, stereophonic sound

Choose a term that you can define without having to consult a dictionary. Try to define it so that the average high-school reader will understand and be interested. Decide which methods of expanding the logical definition will be most useful, and which procedure, deductive or inductive or a combination of the two, will be preferable.

Beneath the title of your paper indicate (1) what reader you have in mind and (2) what procedure you have used. In the left-hand margin identify the methods used. Study the following example on page 64.

A DEFINITION OF A "SEMICOLON"

Reader: an average college freshman

Procedure: inductive

Etymology

History

The word "semicolon" is derived from the Greek prefix semi-, meaning "half," and the Greek root colon, meaning "limb." Later colon came to signify the slow movement of the chorus in a Greek drama from one side of the "orchestra" to the other and back. The "semicolon," then, was the half-way or turning point in the strophe, the dignified dance and song that occasionally interrupted the dialogue of the players.

Comparison

When punctuation was invented, it was logical to use the term "semicolon" to designate the "turning point" in a sentence, and to mark it by using a combination of the period and the comma (;). The period marked the end of a complete statement; the comma marked any minor break in the flow of thought. A semicolon was half way between.

Analysis

In modern usage the semicolon is used for three different purposes:

Example

(1) To separate independent statements not joined by a conjunction, but too closely related in thought to be separated by a period. A typical example is the next-to-the-last sentence in the second paragraph:

> The period marked the end of a complete sentence; the comma marked any minor break in the flow of thought.

A period could be substituted for the semicolon, but the two statements are closely related in thought and are parallel, or "balanced," in structure, facts that can best be indicated to the reader by the semicolon.

Example

Independent clauses joined by a conjunctive adverb, such as therefore and nevertheless, or an adverbial phrase, such as in fact and on the other hand, are similarly punctuated. An example is the following:

> The clauses are independent but related; therefore a semicolon separates them.

Division

(2) To separate items in a series in which one or more of the items has internal punctuation. Here is a typical example:

> The brothers live in different states: John, in Des Moines, Iowa; Bob, in Indianapolis, Indiana; and Bill, in Boise, Idaho.

The two semicolons here are "strong" commas and serve to separate the three items in the series much better than ordinary commas would. If the names, towns, and states here were written with eight commas, the main groups of words would be less easy to distinguish.

Division

Example

(3) To separate main clauses joined by a conjunction, with one or both of the clauses having internal punctuation. Here is an example:

> The Southern states, particularly Mississippi and Alabama, have been reluctant to integrate their public schools; and even the Northern states, especially in the metropolitan areas, have accomplished little more than token integration.

Comparison

Two alternatives are possible here: the "and" after the semicolon could be omitted, and the sentence written as another example of the first usage described above; or the semicolon could be changed to a period, followed by another sentence, either with or without "and."

Division

Another less common use of the semicolon is to separate dependent constructions from main clauses; for example,

Example

> Some men are at their best when mingling with people; shaking hands and talking with them face to face.

Though common in informal writing, this use of the semicolon is not approved by most authorities, who prefer the comma to separate such constructions.

Logical definition

We can now define a semicolon as a punctuation mark not so strong as a period, but stronger than a comma, used to separate main clauses not joined by a conjunction, to separate items in a complicated series, occasionally to separate main clauses joined by a coordinating conjunction, and much less commonly to separate dependent constructions from main clauses.

D. *What methods of expansion, and what procedure does the author have in mind in each of the following definitions? Can you suggest ways of improving any of them?*

THE ZEIGARNIK EFFECT ON MEMORY

As early as 1873, Herbert Spencer recognized that the mind selects what we remember and what we forget. He defined pleasure as "a feeling we seek to bring into consciousness" and pain as "a feeling we seek to get out of consciousness." Early studies based on replies to questionnaires tended to show that there was indeed greater recall of pleasant than of unpleasant experiences.

Legend has it that several decades later, Kurt Lewin and his students were puzzled by an occurrence in a Berlin beer garden. A waiter there could retain long, detailed, complicated orders without writing them down. Once, after the meal had been served and he had given the party their bill, however, someone asked him a simple question about the order. It turned out that he could remember very little of it once he had completed his task.

These findings led to a classic experiment that demonstrated greater recall of tasks before completion than of comparable tasks after completion. In other words, people remember unfinished tasks better than finished ones. This effect of enhanced recall for uncompleted tasks was named the Zeigarnik effect after Bluma Zeigarnik, the student who carried out the study in 1927. In her experiment, the subjects performed simple tasks which they could finish if given enough time, such as writing down a favorite quotation from memory, solving a riddle, molding an animal from clay, and doing mental arithmetic problems. In half of the tasks, the subjects were interrupted before they had a chance to carry out the instructions in full. In the other half of the tasks, they were allowed to finish.

Despite the fact that the subjects spent more time on the completed tasks than on the interrupted ones, they tended to recall the unfinished tasks better when they were questioned a few hours later. Eighty percent of the subjects could name more of the uncompleted tasks than of the completed ones. Within twenty-four hours, however, this better recall of the uncompleted tasks disappeared.

Last week, when I was describing a calculus test to my roommate, the Zeigarnik effect was exemplified. The three problems I didn't get finished, I could remember; but those I had completed escaped my memory. The next day when I was explaining the examination to another friend, I remembered two of the uncompleted problems and two of the ones I had finished.

In attempting to explain this effect, some investigators have postulated that people who are actively working on a task develop mental "task tensions" that do not dissipate until the job has been completed. If it is not completed because of some interruption, these tensions remain. According to the theory, the individual should continue to think about the uncompleted task and should seek to finish it if given the opportunity.

As often happens, more recent experimentation has noted more specific conditions under which the Zeigarnik effect is revealed. For example, when the situation is stressful (threatening, dangerous, or anxiety-provoking), the effect will not occur. Likewise, when the individual is aware that the task measures self-esteem or personal success, the Zeigarnik effect probably will not be observed. Indeed, it may even be reversed; that is, the completed tasks may be remembered better than the uncompleted ones. Interruptions in such cases were presumably interpreted as failures, and the subjects did not want to remember them, or at least to admit remembering them.

In educational or industrial settings the Zeigarnik effect has yet to be carefully studied. The implications for incentives and motivation need more attention than they seem to have had.

—Jacque Moss

ENERGY

Energy is an entity rated as the most fundamental of all physical concepts and usually regarded as the equivalent of or the capacity for doing work either being associated with material bodies (as a coiled spring or speeding train) or having an existence independent of matter (as light or X rays traversing a vacuum), its physical dimensions being the same as those of work $ML^2 \div T^2$ where M is mass, L length, and T time, usually being expressed in work units (as foot-pounds or ergs), and in any form being endowed with the properties of mass (as inertia, momentum, gravitation) by relativity which assigns to the energy E an equivalent mass m by the equation $m = E + c^2$ where c is the speed of light.

—By permission. From *Webster's Third New International Dictionary*, © 1976 by G. & C. Merriam Co., Publishers of the Merriam-Webster Dictionaries.

UNEMPLOYMENT

Unemployment, as the term is used in economics, is a condition in which people who are willing to work, are able to work, and are ordinarily doing work for a wage, do not have jobs. Such people may be grouped in five classes, the names of which indicate why they are out of work.

1. *Residual unemployment* occurs if I quit my job at Monsanto Chemical Company and start looking for other work. I would be so classified only while job hunting. In an economy such as ours, many people are likely to be temporarily in this group.

2. *Seasonal employment* is most common in agriculture between the planting and harvesting of crops. For example, my father used to hire a man in the spring to help put in the crops, and then again in the fall to help harvest them. During the summer and winter, when there was less work on the farm, the hired man would be out of a job.

3. *Technological unemployment* occurs when a new machine or a new process makes a person's job obsolete. In rapidly growing industries this, like residual employment, often happens. Eventually the worker is likely to find a new job.

Of these three types, little can be done to limit or reduce the numbers of people affected. Usually the people will find work soon. The fourth and fifth types are of more importance in our economy, though they may be controlled or regulated to some extent by public and private policies and programs.

4. *Cyclical unemployment* is a result of the speeding-up and slowing-down trends in our economy as a whole, or even in one industry. A good example would be in house-building, which follows a fairly regular three- to four-year pattern, or cycle. In one year many houses will be built; two years later very few will be built; and then in another two years many houses will again be built. In the slack period, carpenters, the people most affected, will be out of work. Carpenters are thus subject to cyclical unemployment. Government policies to stabilize the economy, that is, to stop or reduce the intensity of the cycles, may be helpful.

5. *Structural unemployment* results when people lack the proper skills to fit into the available jobs. Square pegs may be trying to fit into round holes: the round holes are the available jobs, and the square pegs are people without proper skills and training. Continuing education, training programs, and relocation programs are methods of reducing structural unemployment.

Our original sentence definition of unemployment needs some clarification. First, a person is willing to work when he does not refuse to work. Second, if a person is sick or has a broken leg or some other physical disability, he belongs to none of the five categories listed. Third, a person whose only source of income is from stocks, bonds, or other investments and who does not work for a wage, is not regarded as unemployed. People unwilling to work, unable to work, or not working for a wage are not included in an unemployment list.

—Everett Tobiason

Notes

1. *Journal of Home Economics,* October, 1940.
2. By "abstraction ladder" the semanticist means a series of terms ascending from the object itself to increasingly broader and more abstract terms of designating it; for example, from Bessie, the animal itself, to cow, to domestic bovine, to quadruped, to mammal, to animal, etc.
3. By permission. From *Webster's Third New International Dictionary,* © 1976 by G. & C. Merriam Co., Publishers of the Merriam-Webster Dictionaries.
4. By permission. From *Webster's Third New International Dictionary,* © 1976 by G. & C. Merriam Co., Publishers of the Merriam-Webster Dictionaries.

5. Perhaps it should be explained here that these steps in framing a logical definition apply mainly to nouns and noun phrases. Other parts of speech are usually defined by giving synonyms or explaining their meaning in other terms; for example,

define, verb: to fix or mark the limits of
concrete, adjective: naming a real thing or class of things
nevertheless, adverb: in spite of that

No *genus* and *differences* are identifiable in these explanations.
6. "What Do You Mean?", from *Science and the Goals of Man.*
7. By permission. From *Webster's New International Dictionary, Second Edition,* © 1961 by G. & C. Merriam Co., Publishers of the Merriam-Webster Dictionaries.
8. By permission. From *Webster's Third New International Dictionary,* © 1976 by G. & C. Merriam Co., Publishers of the Merriam-Webster Dictionaries.
9. By permission. From *Webster's New Collegiate Dictionary,* © 1980 by G. & C. Merriam Co., Publishers of the Merriam-Webster Dictionaries.

6

Classifications

Classification is the process of arranging any miscellaneous collection of things in groups to make clear their similarities and differences. It is the second part of the process of definition (differentiating an object from other members of its class) applied to all members of the class. For example, to define *table* is (1) to assign it to a class of objects (articles of furniture) and (2) to list the qualities that make it different from other articles of furniture; to classify *tables* is to arrange them in groups according to the ways in which they resemble one another.

IMPORTANCE OF CLASSIFICATION

Consciously or unconsciously you are frequently engaged in classifying things. When you get up in the morning, you classify the weather: cold, hot, rainy, foggy, etc. When you meet people in the street, you classify them: friend, stranger, doctor, tramp, policeman, etc. Such classification is usually unsystematic, informal, and incomplete.

But much of our scientific progress has been due to formal and systematic classification of facts. From one point of view science is "a branch of study which is concerned with observation and classification of facts." Therefore a knowledge of the principles of classification is indispensable to the scientist. Only by classification can anyone reduce the infinite complexity of separate facts to the order that makes possible the formulation of general laws. All of the great advances in the accumulation of knowledge have been due to generalizations based upon systematic classification.

LIMITED AND UNLIMITED CLASSIFICATIONS

Unlimited classifications include all members of a class, for example, *automobiles* (all automobiles made any time, anywhere). Limited classifications include only a portion of the whole class, for example, *automobiles made in the United States in 1975.* Unlimited classifications require comprehensive knowledge of all members of a class and are therefore difficult to make. Compare the problem of classifying *all hand tools* (unlimited) with that of classifying *the hand tools in my shop* (limited).

FORMAL AND INFORMAL CLASSIFICATIONS

The writer of technical papers is concerned, of course, mainly with formal systematic, objective classifications. The statement that houses may be classified as comfortable, inviting, stiff, homey, etc., would have but little value for the architect because it is based on no objective principle: what may be "comfortable" for you may be hopelessly "stiff" for someone else. But the classification of houses according to number of rooms (4-room, 5-room, etc.) or according to material used in the walls (wood, brick, concrete, etc.) is objective and systematic.

REQUIREMENTS FOR FORMAL CLASSIFICATION

The requirements for good formal classification are of two kinds: (1) those preliminary to the making of any classification; (2) those to be applied as tests after the classification has taken form.

Preliminary Requirements

1. *Make sure that the genus is such that its members can be formally classified and that the principle of classification is a practicable one.* It is possible to classify engines and books and fish, but not the forms of life on Mars, or Tom Foster's thoughts about a girlfriend. It is possible to classify a given collection of books according to author, publisher, date and place of publication, and subject (the last with only indifferent success, as anyone who has used a subject index in a library will agree); it is not possible to make an acceptable classification according to quality—great books, mediocre books, bad books. The criteria for a formal classification must be in the objects themselves, not in the minds of the observers. Probably no two people would agree on a list of the ten greatest books.
2. *Have a reasonably accurate and complete knowledge of the things to be classified.* Do not attempt to classify transformers or the fishes of the Caribbean Sea if you lack adequate knowledge—that is, if you intend the classification to be useful. For entertainment or self-education you may amuse yourself, of course, as you like.
3. *State or have clearly understood the principle of classification.* Almost any group of objects can be classified according to many principles. Automobiles can be classified according to manufacturer, style of body, horsepower, wheelbase, price, type of engine, etc.; but any single classification should be according to only one principle.
4. *Adopt a useful principle of classification.* The more generalizations that can be made from a classification, the more useful it will be. A classification of automobiles according to color might be of some value to a paint manufacturer, but a classification according to type of engine or body would be more generally useful.

For a specific purpose, of course, a classification according to even the most limited of principles may be justified.

Tests of a Classification

After you have made the classification, test it by answering the following questions.

1. *Is the same principle of classification adhered to throughout?* You may make several different classifications of the same collection of objects according

to as many different principles. Bridges, for example, may be classified according to *purpose* or *function, length, material, type, method of providing clearance for navigation,* etc. But within any one classification the principle should remain the same.

In subclassifications, however, the principle may be different from that governing the main divisions, especially if it is impossible to subclassify according to the same principle. For example, automobiles classified according to manufacturer (Chrysler, Cadillac, etc.) could be subclassified only according to some other principle, such as body type: sedan, coupe, station wagon, etc.

Whether such a shift of principle in subclassifications is justifiable or desirable depends on the uses for which the classification is intended. Though logicians disapprove such a shift, the resulting classification may be useful.

2. *Is there any overlapping?* It should be impossible to classify any particular member of the whole class under more than one head. A classification of automobiles into Fords, Chevrolets, and trucks would be overlapping because an automobile might be both a Ford and a truck or a Chevrolet and a truck.

Overlapping will ordinarily result when there is a shift in the principle of classification.

Sometimes, however, you cannot avoid some overlapping, either because your knowledge of individual differences is faulty or because there is no sharp, clear line of division between sections. The divisions of a classification of the form of life into animal and vegetable overlap because, as the differences are ordinarily defined, certain living forms might be either animal or vegetable. In a classification of machines by function the divisions may overlap because the same machine may perform different functions. For example, an electric motor may be used to transform electrical energy into mechanical energy, or it may be used to transform mechanical energy into electrical; if it does the latter, it is no longer a motor in function, but a generator. If you attempt to classify people according to intelligence or height or shape of the head or color of the skin, the divisions will shade into each other so gradually that only by making arbitrary distinctions can you avoid overlapping.

3. *Are the items of your outline really divisions or merely descriptive terms?* If a classification of books has a division *hardcover,* with subheads *durable, expensive,* and *heavy,* the subheads are merely descriptive: the adjectives may apply to any member of the class.

4. *Is the classification complete?* If it is complete, every member of the whole class will be included somewhere. A classification of automobiles according to manufacturer would be incomplete, for example, if it did not include Maxwells, Wintons, and Packards—unless the field is limited to "automobiles now being manufactured."

Incompleteness may result from either oversight or lack of knowledge. Any classification of fish made by the most competent of ichthyologists is probably incomplete because some species of fish have not yet been discovered.

You can usually avoid incompleteness by carefully defining and limiting the field of things to be classified. Instead of *automobiles* you can limit the field to *automobiles manufactured in the United States in 1975;* instead of *fish, fish in American freshwater lakes.*

5. *Does every classification or subclassification have at least two divisions?* If it does not, either your grouping is incomplete or the item standing alone is not the result of division. You must distinguish here between *subdivisions* and *lists.* If you are classifying people according to age and one division is "young people," the subhead "babies" would designate a subdivision. The subclassification would be incomplete without other subheads, such as "children" and "teenagers." But if you are classifying bodies of water and one division is "oceans," with the subheads "Pacific" and "Atlantic," the fault is merely that you have an incomplete *list.* There are five oceans. Subheads like "very large" and "salty" would be merely descriptive and quite out of place in a classification outline.

Such a list, supplementing division, may sometimes be desirable in a classification outline. The last step in classification is often to list the particulars in each division, especially if the number is small.

A classification of horse-drawn implements according to use on one northern Indiana farm in 1910 would need four main divisions. Only subheads II A and C, III B, and IV D require subclassifications. Other items are merely lists of individual implements.

I. Implements used before plowing
  A. Roller
  B. Drag
  C. Stoneboat
  D. Manure spreader
II. Implements used in preparing the seedbed
  A. Plows
    Walking
    Riding
  B. Roller
  C. Harrows
    Spring-tooth
    Spike-tooth
    Disk
III. Implements used in caring for the growing crop
  A. Weeder
  B. Cultivators
    Walking
    Riding

IV. Implements used in harvesting
  A. Mower
  B. Binder
  C. Corn-cutter
  D. Rakes
    Dump
    Side-delivery
  E. Baler
  F. Hayloader

Whether such a listing is desirable will depend on the number of particulars and the purposes of the classification.

6. *Is the classification in convenient form for the reader?* If it is merely a classification outline, observe the usual conventions for acceptable outline form:

a. Unless the outline is brief, use symbols to designate the different items, in some systematic order such as the following:

I...
  A...
    1...
      a...
        a′...
        b′...
      b...
    2...
  B. ....
II...

b. Begin coordinate items (A, B, C, etc.; 1, 2, 3, etc.) in the same vertical space.

c. If an item runs over a single line, begin the second and succeeding lines (called "turned-over lines") exactly below the initial capital letter of the first line. (An acceptable alternative is to indent turned-over lines slightly more than first lines of items. Whichever layout you use, be consistent.)

d. Make coordinate items grammatically parallel; for example,

A. Government bonds B. Municipal bonds	not	A. Government bonds B. Bonds issued by municipalities

e. Be consistent in capitalization and punctuation. Alternative usages are permissible: open punctuation omits all periods after all symbols and items in the outline; closed punctuation has periods after all symbols and all items. The most common usage, as illustrated, is to insert periods after symbols, but not after items in the outline.

Initial letters of all important words may be capitalized, but more commonly only the initial letter of the first word of each item is capitalized. Sometimes in complicated outlines initial letters of items designating the lowest levels of classification are lowercase letters.

THE CLASSIFICATION PAPER

Frequently the classification outline—like other forms of outline—serves as the basis for an expository paper in which the items are expanded by giving definitions, explanations, illustrations, etc. For convenience in reading a classification paper, it is helpful to retain the symbols and use the items of the outline as headings.

EXERCISES

A. *What are the faults in the classification outlines below?*

A classification of United States coins according to material:

I. Silver coins
 A. Dollars
 B. Half-dollars
 C. Quarters
 D. Dimes
II. Nickel
 A. Five-cent pieces
III. Copper coins
 A. Pennies
IV. Gold coins
 A. Dollars
 B. Half-eagles
 C. Eagles
 D. Double-eagles

A classification of the students registered in a section of English 414:

I. Major fields of study
 A. Engineering
 B. Science and humanities
 C. Home economics
 D. Agriculture
II. Affiliations
 A. Fraternity
 B. Nonfraternity
 C. Sorority
III. Classes
 A. Junior
 B. Senior
 C. Graduate
IV. Sources of support
 A. Complete scholarship
 B. Partial scholarship
 C. Self-supporting
 D. Partly self-supporting

B. *Select a general class of things with which you are familiar and make two classification outlines, according to two different principles, of the members of that class. Keep in mind the prerequisites and the tests enumerated in this chapter.*

The following limited-classification outlines are satisfactory:

I. Large natural lakes in the United States (over 100 square miles in area) according to area

Lakes 35,000-1,000 square miles in area

Superior (31,800)
Michigan (22,400)
Erie (9,910)
Huron (9,100 in U.S.)
Ontario (3,460 in U.S.)
Great Salt (1,500)
Iliamna (1,033)

Lakes 1,000-500 square miles in area

Okeechobee (700)
Pontchartrain (630)

Lakes 500-200 square miles in area

Red (451)
Champlain (430)
Salton (360)
Winnebago (315)

Lakes 200-100 square miles in area

Tahoe (192)
Pyramid (187)
Upper Klamath (140)
Yellowstone (137)
Bear (136)
Pend Oreille (133)
Moosehead (117)

II. Large natural lakes in the United States (over 100 square miles in area) according to surface altitudes

Lakes over 5,000 feet in altitude

Yellowstone (7,735)
Tahoe (6,229)
Bear (5,930)

Lakes 5,000-3,000 feet in altitude

Great Salt (4,200)
Upper Klamath (4,139)
Pyramid (3,800)

Lakes 3,000-1,000 feet in altitude

Pend Oreille (2,063)
Moosehead (1,028)

Lakes 1,000-200 feet in altitude

Superior (600)
Michigan (576)
Erie (570)
Huron (518)
Red (451)
Champlain (430)
Winnebago (215)
Ontario (214)

Lakes 200-0 feet in altitude

Iliamna (50)
Okeechobee (19)
Pontchartrain (0)

Lakes below sea level

Salton (−231)

C. *Using one of the outlines prepared for Exercise B, write a classification paper in which you define and expand the items of the outline. Use headings to mark the divisions of the classification. The following are simple examples of such a paper.*

A CLASSIFICATION OF TYPES OF SOIL

A limited classification of types of soil is in accordance with the *parent material,* consolidated mineral matter and unconsolidated mineral matter, the layer from which soil is formed. The chemical and mineralogical characteristics of the parent materials determine the type of soil formed and the rate of soil formation. Soil types are determined by the *agency of deposition* of the parent material; that is the force or lack of force that results in deposits of parent material in quantities large enough to allow the formation of soil.

According to this principle two types of soil can be distinguished, sedentary parent material and transported material.

I. *Sedentary Parent Material*

Sedentary parent material is the unconsolidated mineral matter that has accumulated from the bedrock it lies upon. As "sedentary" indicates, the parent material has not moved from the point of its formation.

Only one type of soil, *residual soil,* is produced from sedentary parent material. It has not moved from the place where it was originally formed.

II. *Transported Parent Material*

Soils made of materials that have been moved by the force of gravity, water, ice, or wind are termed *transported.* Six subclassifications can be distinguished:

A. *Colluvial Soils*

Colluvial soils are produced from parent materials that are carried from higher to lower elevations by gravity.

B. *Alluvial Soils*

Alluvial soils are produced from parent materials deposited by moving water.

C. *Marine-Type Soils*

Marine-type soils are produced from sediments that have been exposed to weathering by the recession of the sea. In this sense "marine" indicates that the agency of deposition of the parent material was salt water.

D. *Lacustrine Soils*

Lacustrine soils developed from parent material deposited in glacial lakes during the Ice Age.

E. *Glacial Soils*

Glacial soils were formed from parent materials deposited by the action of glaciers.

F. *Loessial Soils*

Loessial soils have resulted from the material deposited by the movement of material by the wind.

III. *Value of the Classification*

A classification of types of soil according to the soil-forming factors of kinds of parent materials is of primary value in the historical study of soils. From the study of soil types and their relationship to their parent material(s), the path of a glacier can be followed or the site of an ancient lake or ocean can be determined. The formation of continents, the development of different cultures, and even much of the history of mankind have depended largely on the relationships of soil types and parent materials.[1]

—James Thompson

CLASSIFICATION OF DISORDERS OF ARTICULATION

Disorders in articulation—faults in the production of speech sounds—are the most common of speech defects. There are two kinds of such disorders, organic and functional. The organic problems are caused by biological defects, such as cleft palate, hearing loss, and deformities of the mouth or teeth. This paper is about the other kind, functional, or non-organic, articulation disorders. They can be classified according to the type of phonetic error. The speech therapist analyzes the speech of a client and determines whether the errors are due to omission, substitution, or distortion.

Omission.—Omission is the term used to designate speech in which a sound is not produced. Pronunciations such as "ook" for "look," "ba—" for "ball," and "p—ay" for "play" are illustrations of this disorder. Omissions will most commonly involve initial or final consonants, such as in "—ook" and "ba—," or an element of a consonant blend as in "p—ay." They are characteristic of infantile and immature speech. The child has not yet developed the auditory discrimination to hear these sounds in the speech of others. If he is still omitting them when he reaches school age, he should receive speech therapy.

Substitution.—Substitution is the disorder in which one standard speech sound is substituted for another. For example, a child may say "wed" for "red," "Tudie" for "Suzie," or "thit" for "sit." Sounds that are acquired early in language development, such as (w), (d), (t), (p), and (b), are substituted for sounds that are usually acquired later, such as (r), (s), (z), (l), and consonant blends. The child has developed enough auditory discrimination to recognize the presence of a sound, but he is unable to differentiate between the right and wrong sounds. If he has this type of articulation problem, he should be referred to a speech therapist.

Distortion.—Distortion, like substitution, replaces one sound by another; however, the sound substituted is not a standard speech sound. Speech in which the *s* is replaced by a sound resembling an *s* with a whistle is a distortion, caused by unrefined auditory discrimination. The child makes a close approximation to the desired sound, but does not hear the difference. Therefore he may learn to make the sound incorrectly. If distortions are noticed in the speech of a child eight years old or older, he needs the attention of a speech therapist.

Summary.—These types of problems in articulation are related to each other in the development of a child's speech. As he acquires the ability to speak, he may have all three types. When any of the types persists after certain ages, the child needs specialized speech instruction. In order to be corrected, all three require refinement of auditory discrimination.

Omissions are the earliest types of disorder to be observed. They are also the most serious because they are the least intelligible. Substitutions are second in order of development. They are often replaced by distortions, which are last in order of development. A child may exhibit all three of these types of disorders at the same time; on the other hand, he may never make errors in articulation. If he does, he may eventually outgrow them.

References

Carrell, James A., *Disorders of Articulation*
Millsapps, John, Instructor, Iowa State University, Speech 376
Vallier, Fred, Instructor, Iowa State University, Speech 375
Van Riper, Charles, *Speech Correction*

—Nancy Jones

Notes

1. D. E. Green and D. G. Moolley, *Audio-Tutorial Biology Series in Agronomy*, Burgess Publishing Company, 1969.

7

Analyses

Analysis, the opposite of synthesis, consists of resolving a subject into its elements, or basic parts. It is derived from Greek roots *ana,* up, and *lyein,* loosen.

SOME DISTINCTIONS

Two of its commonest types are *physical* and *logical.* Physical analysis, which includes chemical, breaks a material object or substance into its elements; logical analysis breaks an idea or concept into its elements. In many ways the chemical and physical processes are like the logical, but logical analysis is a product of the mind rather than of test tubes and retorts.

Difference Between Analysis and Classification

Comparing analysis and classification may be helpful. In classification you begin with particulars, each a part of the whole subject. Your task is to group the particulars according to the similarities and differences you may find. In analysis, however, you begin with a whole and break it down into parts. Then you group and arrange the parts by the methods of classification.

A report on the machines in a factory might call for a classification—a grouping and listing of the machines according to function or cost or age. A report on the processes carried out in a factory would more likely require analysis—breaking the process into its parts. In the classification many particulars are to be grouped. In the analysis a whole is to be divided into its parts.

Since classification and analysis often seem to bring identical results, they are easily confused. A classification of the people living in X-town might look exactly like an analysis of the town's population. But the classification would begin with the individuals who live there, probably with a census for collecting data, then a systematizing of the data, perhaps with a computer, ending with grouping according to some principle of classification such as occupation, health, race, income, or age. Analysis, however, would begin with the whole population and would group the people according to a predetermined principle.

Classification, then, is concerned with particulars and begins with them. Analysis is concerned with the whole and begins with it.

Analogy Between Analysis and Factoring

Factoring in mathematics is like analysis in that it begins with a number that can be broken down into smaller numbers whose product is the original number. For example, the figure 48 can be resolved into factors 6 and 8, and these further resolved into the prime factors 3, 2, 2, 2, 2. In logical analysis the word "factor" is often used to designate an element determined by analysis. Instead of the whole being a "product" of the factors, it is more accurately regarded as the "sum" of the parts. If we let U stand for the original unit and *a, b,* and *c* for the parts identified by analysis, the formula would be

$$U = a + b + c$$

rather than

$$U = abc.$$

Diagnosis a Combination of Classification and Analysis

The medical diagnosis is a good example of a combination of classification and analysis. The nurse observes and records symptoms: the patient has a temperature of 104 degrees, a pulse of 100, a pain in the chest, a rasping cough, etc. But the nurse's data are the evidence that more than classification is necessary. The diagnostician must perform an analysis to determine how the symptoms relate to the patient's ailment.

Formal (Scientific) and Informal Analysis

An analysis that is objective, exact, and complete is properly called *formal* or *scientific*. One that is subjective, approximate, and incomplete is called *informal.* An analysis of the enrollment of students in a college indicating where their homes are or what grades they are making would be formal because the data are specific. An analysis of the character of the typical Irishman would be informal because it is not subject to exact objective treatment. With the first subject any two careful analysts would arrive at the same divisions; with the second they would not.

Between two such extremes lies a wide range of subjects. Many of them cannot be analyzed with complete formality because our knowledge of them is incomplete. The best informed economists still disagree on the causes of and the remedies for inflation. Experienced psychologists will disagree on an analysis of the kinds of thinking. Such subjects are still more suitable for formal analysis, however, than, say, the effects of music on the human soul, or the relative merit of the prose styles of Mailer and McPhee.

In general, formal analysis uses the language of scientific prose, informal analysis is personal and impressionistic. It is possible, of course, to analyze the effects of seeing a sunset in factual and dispassionate language, and on the other hand to interpret statistics concerning poverty in America in flaming and explosive terms, but in general the more complete and exact the information about a subject, the more calm, impersonal, and unprejudiced the analysis of it is likely to be.

SOME ILLUSTRATIONS OF TYPES OF ANALYSIS

The variety of kinds of logical analysis can be shown by listing a few types of subjects suitable for analysis:

- Describing mechanisms and processes (analysis is involved in breaking the whole into its parts or steps)
- Finding the causes or the effects of a phenomenon
- Demonstrating the faults or merits of a proposal
- Determining the feasibility of a project
- Specifying the requirements for a job or undertaking
- Making time and motion studies and job analyses
- Ascertaining the status of a partially completed program
- Estimating costs and depreciation, future plant needs, etc.
- Comparing the relative merits of machines, methods, etc.
- Reducing a problem to its elements
- Making an organization chart
- Describing a factory layout
- Making a proposal to managers and indicating its merits and faults

REQUIREMENTS FOR SCIENTIFIC ANALYSIS[1]

Since the use of the scientific method requires constant formal analysis, it is evident that, if you are the analyst and anything more than mere mechanical manipulation is involved, you will need the same qualities as the good scientist: knowledge of the field, complete honesty, freedom from prejudice, good judgment, and patience.

Knowledge of the Field

Since your problem is to break a subject down into its elements, it is obvious that the more you know about the field in which the subject lies, the more successful you will be. If you knew nothing about chemistry you could not analyze a compound like sulfonmethane. If you set out to analyze the faults of the public school system in the United States, you should have had a long experience as a public school teacher or administrator.

Complete Honesty

You must be honest if your results are to be useful to others. The report of the auditor who analyzes the financial condition of a bank must be free of any suspicion of dishonesty if other people's confidence in the bank is to be maintained. The auditor must be completely disinterested, neither an officer of the bank nor a stockholder in it.

Freedom from Prejudice

You may be free of all suspicion of dishonesty and yet unconsciously permit a prejudice to influence your analysis. Students analyzing the effects of a new rule prohibiting the parking of their cars on the campus might be likely to ignore or belittle the importance of many factors that might affect their own convenience. The man who owns oil wells in Texas might well be unable to render an unbiased report on the merits of federal oil-depletion allowances. Since prejudices are often unconscious or unadmitted, they affect the reliability of many analyses. Even in pure science the preference of a worker for a theory that he or she has conceived may blind the worker to the merits of every other theory. Lincoln Steffens relates that a brilliant young graduate student working for a degree under the great German psychologist Wundt deliberately falsified the results of some research work because he knew that Wundt was prejudiced against the theory that the student's data seemed to support.

Good Judgment

Though good judgment is necessary in any kind of writing, it is especially necessary in analysis where causes and effects and faults and merits must be determined and generalizations must be made. Determining the comparative values of certain devices or methods or policies calls for the highest type of good judgment. Where should the new freeway be located? Should the new research center be located in the East or the Midwest? Should X-company install a quality-control system? Decisions on such questions can be reached only after exhaustive analysis of innumerable factors and careful weighing of alternatives. Engineers must constantly be deciding whether a proposed construction will pay, what safety factors must be allowed for, why a bridge collapsed, how a strike can be averted, where a dam should be built. All such questions demand the exercise of good judgment in analysis.

Patience

Patience is necessary for the formulation of good judgments. If you are impatient you "jump to conclusions," which means that you reach conclusions before you finish your analysis. Even with all the evidence before you, you will find it difficult to make dependable judgments. If you reach conclusions too soon, you are likely to be wrong.

ORGANIZATION OF THE ANALYTICAL PAPER

In writing an analytical paper you need first to make an analytical outline, presenting in systematic form the parts or elements of the whole subject. Follow the same principles that govern the making of a classification outline.

If (to take a simple example) you are making a thorough analysis of the condition of a gasoline engine, the proper procedure would be to tear the engine down and lay out all its parts in convenient order: in one place the cylinder block, the head, the head bolts, etc.; in another the piston, piston rings, connecting rods, etc.; in another the carburetor; and so on. The report of the results of the analysis you would organize in the same fashion, with a division and a suitable heading for each part of the engine, plus suitable introductory and concluding sections.

If you are analyzing students' attitudes toward voting, the problem is more difficult. At first you would probably devise a questionnaire. Then you would question many students, trying to find typical attitudes—sampling. You would form categories—the politically minded, the indifferent, the ignorant, the prejudiced, the scornful, the satirically inclined—some probably with subclassifications. The result would be an outline for a classification paper, which would call for definitions, examples, and speculations about causes and effects.

RULES FOR MAKING AN ANALYTICAL OUTLINE

Since the analytical outline is similar in its final form to a classification outline, the same rules govern the making of both. These rules are listed and explained in the chapter on "Classification." They are repeated here, with illustrations of their application from the field of analysis.

Preliminary Requirements

Some of the preliminary requirements listed below are particularly important in objective scientific analyses.

1. *Choose a subject that you can analyze objectively and scientifically.* The Schoolmen of the Middle Ages made an exhaustive analysis of the "celestial hierarchy," dividing it into three groups with three kinds of heavenly beings in each, ranging from the seraphim to the angels; and the only bases for the whole analysis were a few brief Scriptural references. With even less objective evidence ambitious preachers sermonize on God's plans for the universe. Such analysis is imaginative and fanciful, not specific. A well informed person might analyze the plans of a Democratic administration, but the plans of God are beyond human knowledge.

2. *Adopt a single principle of division.* This rule applies to physical and chemical as well as to logical analyses. The division of an ear of corn into cob, grain, and starch would violate the principle because cob and grain are physical

elements and starch is a chemical element. Once the physical elements are determined, however, each can be resolved into chemical elements. When you employ a different principle in subdivisions, you should inform your reader of the shift.

Of course a single subject can be analyzed in different ways. A bank can be analyzed to determine its financial condition, the physical condition of its property, its reputation among its customers, its investment of funds, its history; but the analyses according to these various principles would have to proceed separately.

3. *Have an adequate knowledge of the subject.* Neither chemists nor geologists are likely to be qualified to analyze economic conditions. Too frequently it is assumed that people competent in one field are competent in others: they speak or write oracularly on subjects about which they know little. In many fields qualified scholars will disagree on their own fields of specialization; ignorant people had better hold their tongues.

4. *Let the principle of division be determined by the purpose for which the analysis is made.* An analysis of the topography of a region made for the Department of Agriculture would probably be a division of the area according to the uses for which the land is best adapted; an analysis made for a travel agency would probably be according to the recreational interests of the vacationer.

Tests of the Completed Analysis

Once you have completed the division of your subject, test it in the same ways that you test a classification outline.

1. *Is there a shift in the principle of division?* Though you may make different analyses of the same subject according to different principles, you should not shift the principle within any one analysis (except sometimes, as already noted, in subdivisions).

2. *Is there any overlapping of divisions?* An analysis of the population of a city should not have as one division *clerks* and as another *white-collar workers,* because the same individual might then belong to two divisions. In dealing with abstractions, however, you might find it difficult to draw clear lines of separation. *Love of money* and *love of power* can both be listed as causes of war; it is doubtful if they can be defined so that they will not overlap.

3. *Is the analysis complete?* If it is, you have listed all the possible divisions. A complete analysis of the sources of income of a corporation would have to cover not merely the main sources but all sources, even though insignificant. You can simplify the problem, of course, by lumping together small items under the heading "Miscellaneous." It is quite proper also to make the analysis intentionally incomplete, and indicate the intention by use of such a title as "The Principal Sources of Income."

4. *Is there a real division of the subject?* Since analysis is the division of a whole into its parts, every analysis will have at least two divisions; every analysis of a division will have at least two subdivisions. But to put a subhead "informative" under a main head "Encyclopedias" is not to divide at all, and to put under a main head "College students" two subheads, "two-armed" and "two-legged," is again not division but a listing of characteristics.

5. *Are the divisions of the analysis in logical order?* If you are analyzing a group of people according to religious affiliations, arrange the divisions according to some principle, probably size—the largest group first, the smallest last. If the principle of arrangement is not immediately evident ,tell your reader what it is.

6. *Is the analysis in convenient form for the reader?* An outline enables the reader to see at a glance the whole scheme of the analysis. If you use the items of the outline as headings in the analytical paper, the analysis will be much easier to read.

TYPICAL PROCEDURE IN ANALYSIS

Let us assume that you are employed by a bank which makes real estate loans, and that you have been asked to appraise the value of a house and lot and write a detailed report of your appraisal. Your problem is basically one of analysis since you begin with a unit—the house and lot—whose total value you must estimate. To arrive at an estimate, however, you must consider many details that make up the whole.

In collecting data for your appraisal and report, you carry a pad and pencil with you to record details that may affect your final result. Probably your notes will contain material on the following points:

Age of house
Size of lot
General location
Street improvements
Style of architecture
Materials of construction
Number of stories
Landscaping
Garden
Fences
Porches
Windows
Chimneys
The lawn
Number of rooms
Floors
Heating system
Plastering
Sidewalks
Garage
Access to schools and churches
Bus lines
Roof
Foundation
Arrangement of rooms
Kitchen equipment
Bathroom
Attic
Closets
Woodwork
Lighting fixtures
Radiators
Storm windows
Insulation
Exterior paint
Basement floor and walls

These items are in haphazard order, just as they occur to you during your inspection. For your report you will need to rearrange them, group them according to some logical principle—perhaps *location*. Your main headings may then be

I. General location
II. The lot
III. The buildings
  A. The house
    1. Exterior
    2. Interior
  B. The garage

Of course it is possible to select a principle of division at the beginning and group details accordingly as you think of them. Often, however, the right principle may not be apparent until you have collected most of your data.

With the main divisions established, you can now group your original notes in their proper places. Under "General location" you place "Access to schools and churches," "Bus lines," and "Street improvements." Under "The lot," you place "Size of lot," "The lawn," "Garden,' "Fences, and "Landscaping." You continue grouping until every detail has been properly subordinated.

Problems will come up. Will you list "Windows" under "Exterior" or "Interior"? When you arrange items under "Interior," will you continue to group according to position–Basement, First floor, Second floor, Attic? If so, you will have to take up some points more than once–––Heating system, Lighting fixtures, Plastering, etc. Or will you abandon the principle of location now and group according to functional units–Heating system, Plumbing, Wiring, Plastering, Woodwork, etc.? You will need to compare different possible arrangements and choose the one that presents fewest problems.

By some such process you will eventually arrive at a logical and coherent arrangement of your notes. Only then will you be able to write your report. If, while writing, you think of items omitted, or of a more feasible and convenient principle of division, expand or rearrange your outline. An outline properly used is a guide and servant, not a master.

USE OF HEADINGS

Refer to the sections on the uses, values, and forms of headings in chapter 4.

Use a topical heading at the beginning of each division of the analysis. Such headings are common in technical writing and are becoming increasingly common in all scientific writing.

An introductory paragraph, however, at the beginning of the analysis or of a main section needs no heading. The reader will understand its relation to what follows without its being headed "Introduction."

In an involved analysis several degrees of importance in headings may be indicated. Note the use of headings and subheadings in this book. Almost any textbook or technical journal or report will furnish further examples.

EXERCISES

A. *Choose a suitable topic of which your knowledge is adequate and make an analysis in outline form. Then, using the items of the outline as headings, write the analysis.*

Subjects for formal analysis:

- Composition of a chemical compound
- Soil of a farm
- Topography of an area
- A food
- Contents of a dictionary
- Condition of an automoile, house, heating plant, steam engine, laboratory, bridge, drainage system, etc.
- The causes of erosion
- A job in a factory (motion study)
- The way you spend your time for one week
- A week of local weather
- A college course
- Causes of traffic accidents
- Provisions for safety in a laboratory, factory, mine, etc.

Subjects for informal analysis:

- Editorial policy of a magazine or newspaper
- Your own writing (for length and type of sentences, use of participial phrases, subordination of details, etc.)
- Your speech
- Values of television in education
- Effects of raising income taxes
- Requirements for the ideal personnel director, salesman, research chemist, politician, architect, accountant, etc.
- Requirements for the ideal home shop, testing laboratory, kitchen, fountain pen, etc.
- Parking problems on the campus
- Research in your field

B. *Study the organization, the use of headings, and the first and last paragraphs of the following analyses*:

Notes

1. This section is a fairly typical example of informal analysis.

THE HARKNESS PARK RECREATIONAL AREA

1000 Lincoln Way
Ames, Iowa
October 24, 197_

Iowa State Legislature
Statehouse
Des Moines, Iowa

Gentlemen:

As you requested on August 30, the Forestry and Recreation Commission has studied the possibilities of making a recreational park around Harkness Peak in the Grand Neige Mountains. Below are the findings of the study and the suggestions for making the area more attractive for summer and winter sports enthusiasts.

Present Topography

Harkness Peak is located near the geographical center of the study area.[1] On the summit are a large resort hotel, two meteorological stations, two privately owned radio and television transmitters, and a cafe operated by owners of a cable lift. The summit is accessible from one side by a paved road that follows an old 1895 carriage road, from the other side by a cable lift, and from all sides by hiking trails. Two of the trails have mountain club huts near them, about half way from the base of the peak to the summit.

To the west of Harkness Peak is Jones Peak. Its elevation is 6,000 feet, compared to the 6,887 feet of Harkness Peak. Jones Peak has no access road to the summit, but there is one hiking trail. The base of Jones Peak is heavily wooded, and has not been put to any use.

In the vicinity of the peaks the range is flanked by scenic passes, Lawford on the north and Dedham on the south. Paved highways wind through both of them, and each highway has a stream running beside it. The stream in Dedham Pass has many waterfalls, but tourists cannot stop to look at them because there are no places to park. At the east end of the pass is a small lake used for fishing and boating. Approximately midway through Dedham Pass is a commercialized

[1]To save space, Figure 1, a topographical map of the area, is omitted.

animal farm. In Lawford Pass are no waterfalls. At the west end, leading into the pass, is Ackerman Ravine. This is a snowbowl, and draws early spring visitors to the slopes above the timberline for late-season skiing. It has a narrow headwall and is not considered safe for novice skiers. Connected to the highway through Lawford Pass is the old carriage road that leads to the summit of Harkness. Just after the road leaves the highway, it passes the old Holley House, an historical landmark. The house is in poor condition, but could be a tourist attraction if it were restored. Further up this road near the base of Harkness are two resort hotels, both of moderate size.

Present Recreational Facilities

Fishermen can fish in the streams and the lake, but the lake is too small for boats. Swimming is not common, because the temperatures of the water are low even in summer months.

Skiers now have only one lift, and accommodations for lodging are limited.

Parking for sightseers is limited, especially near observation vantage points. Restaurants that are not crowded are hard to find, since the one on Harkness Peak can seat only sixty.

Present Ownership

About 90 per cent of the 25 square miles studied is owned by the United States and administered by the Department of Agriculture's Forest Service as a national forest. Half the remainder is privately owned, belonging to the three resort motels and including a privately owned parcel of 60 acres on the summit of Harkness Peak. The remaining 5 per cent belongs to the state, including the rights-of-way of the mountain roads and the land rented to the animal farm.

Suggestions

The following suggestions for improving the area are derived from the preceding analysis:

1. It is suggested that the stream in Lawford Pass be dammed at the east end. This would greatly increase the water recreation area, and also make the park more scenic. Boating, too, would then be possible.

2. It is suggested that parking areas be built along the highway in Dedham Pass, near the waterfalls, to make it possible for people to stop and walk around these areas. Since this land is already owned by the state, parking places could be built easily and cheaply.

3. It is suggested that the 60 acres of privately owned land on Harkness Peak be acquired, and developed with observation platforms, restaurants, and parking areas. This would alleviate the crowded conditions due to inadequate facilities and attract more tourists to the park.

4. It is suggested that a ski lift, an access road, and a lodge be built on Jones Peak, and also that a ski run be cut through the heavily wooded areas at the bottom of the peak. These improvements would give beginning skiers an area where they could learn to ski, and the new facilities would attract them away from the more dangerous slopes.

5. It is recommended that a ski lift not be put in Ackerman Ravine unless the area can be fully supervised. Otherwise, it would be too easy for beginning skiers to get to the dangerous headwall.

6. It is recommended that the old Holley House be restored and made a tourist attraction.

Respectfully submitted,

(penned signature)

John E. Nimmo
Commissioner of Forestry
and Recreation

THE PROBLEM OF THE ARGOSY CHEMICAL COMPANY

University Village
Ames, Iowa 50010
January 20, 197_

The Argosy Chemical Company
3415 South State Street
Chicago, Illinois 50424

Attention: Mr. Norman Bruton, President

Gentlemen:

As you requested on January 6, I am submitting this analysis of the desirability of creating a new staff position of Vice-President of Planning and Organization.

The Problem

Your problem is to decide whether or not the budget function should be moved from the Controller's office to a newly created position of Vice-President of Planning and Organization. The purpose would be to correct a declining trend in net profits in your firm in the past few years.

Previous Situation

Before you hired Mr. Johnson, your staff consisted of the following officers: vice-presidents of Sales, Manufacturing, Industrial Relations, and Legal Affairs, and a Treasurer and Controller.

Sales in 1954 totaled $148 million, with net profits of $2 million. They represented approximately a 0.5% decrease of net profits in one year, and were consistent with the general trend of declining profits over the past several years.

Method

To determine if the proposed addition to your staff will solve your problem, it was necessary to analyze your past financial statements, including present and past corporate budgets. The financial ratios, summarized in Table 1-1,[1] and an example of your last budget, shown in Figure 1-2,[1]

[1] To save space, not included in this report.

give a good indication that your problems have been originating in the Controller's office. High net profits are largely determined by a company's policies in budgeting.

The Budgeting Function

The corporate budget is merely a device for formulating a firm's plans and for exercising control over the various departments. The budget requires a set of performance standards, or targets, which, when used effectively, provide a continual assessment of the results of all facets of an operation. Aggressive but realistic standards are imperative for the economic growth of a company. Failure to set attainable goals can result in declining net profits.

Opinion

Which staff position should have the responsibility for budgeting is purely academic. The only requirements are that it be directed by an upper-level executive, preferably a vice-president, and that he be highly qualified for the job.

It is the general consensus of your staff officers that the proposed solution would be superfluous, and that his function would overlap previously well-defined responsibilities of the Controller.

Although I agree that the proposed position is unnecessary and would only add the cost of an additional salary, the problem does not involve any overlapping of duties. Your declining net profits are a direct consequence of the inferior performance of your controller, Mr. Thomas.

It is beyond the scope of this report to elaborate on the inefficiencies of Mr. Thomas, except to say that he lacks a good understanding of the budgeting procedure. An examination of procedures in the past revealed the inability of Mr. Thomas to assess properly the capabilities of your firm. His unrealistic fixed and variable budgets have tended to depress the net profits of your firm for the past several years.

Conclusions

Budgeting is a management tool used both for planning and for control. If the budgeting process is competently organized and directed, it can provide valuable guides for both high-level and middle-management executives.

It is unnecessary to incur the additional expense of a high salary for a vice-president of planning and organization when the budgeting function can be effectively performed by a competent controller. Good profit showings are a direct result of a firm's budgetary policies. Unrealistic goals can result in a low morale throughout the organization, and will result in inferior performance on all levels.

What your company needs is not another vice-president, but rather more effective assignment of duties to the present officers.

Recommendations

It is recommended:

1. That you not appoint a new vice-president of planning and organization.
2. That you remove Mr. Thomas as controller and replace him with Mr. Johnson.

Respectfully submitted,

(penned signature)

Alan J. Nelson
Management Consultant

8

Comparisons

One special type of analysis is the comparison of the merits and faults of two or more devices, methods, processes, proposals, etc. It is more complex than the analysis of a single subject, because it covers at least twice as much ground and concludes with a judgment concerning relative merits. For example, a description of the construction and operation of a pencil sharpener is a fairly simple form of analysis; but a comparison of the construction and operation of two (or more) pencil sharpeners is a much more complicated problem, especially when a decision concerning relative merits must be reached.

DIFFERENCE BETWEEN THE "RELATIVE MERITS" PAPER AND PERSUASION

Just as a simple analysis may be scientific or unscientific, so may a comparison of relative merits. When cigarette manufacturers advertise data purporting to prove that one cigarette is milder or has less nicotine than others, even though the tests were made by "eminent medical authorities," the reader is justified in questioning the accuracy of the judgment. It was probably not a scientific comparison. Only a completely disinterested agency whose experts were without bias would be qualified to make a scientific report on such a subject.

In nonscientific prose the comparison of merits for the purpose of inducing the reader to adopt the author's view is called persuasion. Though it may be largely factual, persuasion often goes beyond scientific prose in that it distorts or "slants" the facts and uses rhetorical devices to induce the reader to adopt a certain view. "Persuaders" may fail to mention or may minimize the faults of a proposal; they may magnify the faults of the alternative proposal; they may work upon the reader's emotions: they may state facts in such a way that the reader interprets them wrongly. Authors of scientific comparisons cannot use such methods. They must tell "the truth, the whole truth, and nothing but the truth." They can only present the facts and draw such conclusions from them as a completely impersonal and impartial thinking machine might draw. If the facts warrant no conclusion, no clear preference for either alternative, they must so state.

Good examples of comparative analysis are to be found in the reports of the United States Bureau of Standards, the Underwriters' Laboratory, and the several university experiment stations. Investigators in these agencies make a great variety of comparative tests (on stokers, cotton sheeting, motor oils, tractors, insulating materials, conservation methods, etc.) and report the results impartially and without prejudice. They include all the pertinent data, so arranged that readers can draw their own conclusions or substantiate the conclusions of the authors.

QUALITIES OF A GOOD SCIENTIFIC COMPARISON

The good scientific comparison will have the following qualities.

Freedom from Suspicion of Prejudice

If without reasons or evidence you favor one of the things[1] being compared, or if you falsify data or misrepresent the facts in any way, your paper cannot be called scientific. If scientists are instructed to reach certain conclusions in their "research," they will find it difficult to be impartial and disinterested.

Completeness

The comparison must include all the pertinent data on the things being compared; that is, all that have any bearing upon the conclusions reached. It is quite legitimate, of course, to compare things in certain respects only and to include only the pertinent data. For example, you may compare the fuel consumption of gasoline engines without including data on other factors, such as cost, durability, weight, and construction. If you limit the comparison, however, you should specify what the limitations are.

Clear Organization

If the arrangement is not systematic and clear, your comparison may be complete and free from any evidence of prejudice, but still be poor. If you proceed without a plan, your paper is likely to be confused and jumbled, especially if the subjects are complex or if many points of comparison are involved. Use headings and subheadings to make the plan of organization clear.

PLANS FOR A COMPARISON OF RELATIVE MERITS

Two plans for comparing the relative merits of things are possible:

1. Listing and explaining the merits and faults of each thing separately, first A, then B (perhaps C, D, etc.).
2. Comparing A and B (perhaps C, D, etc.) point by point, both for merits and for faults.

Generalized outlines for these plans (limited here to A and B) are given below. The introduction and the conclusion will be substantially the same for both plans.

I. Introduction (for both plans)
   A. Purpose of the comparison
   B. Description of the things to be compared
   C. Statement of the plan to be followed

II. Comparative data (Plan 1)
   A. First thing
      1. Merits
         a. Merit 1
         b. Merit 2, etc.
      2. Faults
         a. Fault 1
         b. Fault 2, etc.
   B. Second thing
      1. Merits
         a. Merit 1
         b. Merit 2, etc.
      2. Faults
         a. Fault 1
         b. Fault 2, etc.

II. Comparative data (Plan 2)
   A. First point of comparison
      1. First thing
      2. Second thing
   B. Second point of comparison
      1. First thing
      2. Second thing
   C. Third point of comparison
      1. First thing
      2. Second thing, etc.

III. Conclusion (for both plans)
   A. Recapitulation and summary of data
   B. Conclusions

By adding the necessary sections, you can use the same plan for the comparison of three or more subjects.

ILLUSTRATION OF THE TWO PLANS APPLIED

The following outlines of a paper comparing the relative merits of the fountain pen and the ball-point pen illustrate how the general plans described above can be adapted to a specific problem of comparison:

PLAN 1

I. Introduction
  A. The purpose of this paper is to compare the merits and faults of the ball-point pen (A) and the fountain pen (B).
  B. Pens A and B differ in principle and construction.
    1. Pen A has a replaceable inner unit with a ball bearing in the point and a long-lasting ink reservoir filled when made.
    2. Pen B has a standard gold point, with an ink reservoir above that may be filled from a bottle of ink.
  C. In this comparison Pens A and B will be examined separately, first for merits and then for faults.

II. Comparative Data
  A. Pen A has both merits and faults.
    1. Its merits are that
      a. Its first cost is low.
      b. Its point and ink reservoir are replaceable at small cost.
      c. Its ink supply will last a long time.
      d. It rarely floods.
    2. Its faults are that
      a. The width of its mark cannot be varied.
      b. Its ink is usually indelible and difficult to erase.
      c. The ball point sometimes sticks and fails to write.
  B. Pen B also has both merits and faults.
    1. Its merits are that
      a. It may be easily filled from any ink bottle.
      b. It can be easily cleaned or repaired by the user.

PLAN 2

I. Introduction
  (A and B the same as in Plan 1)
  C. In this comparison Pens A and B will be examined in turn for cost, durability, and convenience.

II. Comparative Data
  A. Cost
    1. Pen A is cheap.
      a. Its first cost is low.
      b. The inner unit can be replaced at small cost.
    2. Pen B is relatively more expensive.
      a. Its first cost is higher.
      b. Its pen and reservoir rarely need replacement.
  B. Durability
    1. Pen A is likely to be short-lived.
      a. The ball may become rough or worn.
      b. The point cannot be repaired.
      c. It is likely to be thrown away when its ink is used up.
    2. Pen B will last longer.
      a. Its gold point does not corrode.
      b. Its reservoir lasts indefinitely.

c. It permits shading its mark from thin to thick.
2. Its faults are that
a. Its first cost is more.
b. Its supply of ink is limited.
c. Its writing needs to be blotted.

C. Convenience
1. Pen A is convenient.
a. It will write at once.
b. It is unlikely to flood.
c. It is available with different colors of ink.
d. The writing requires no blotting.
2. Pen A is inconvenient.
a. It cannot be refilled from an inkwell.
b. Its ball point sometimes refuses to function.
c. The width of its mark cannot be varied.
3. Pen B is convenient.
a. It may be easily refilled from any inkwell.
b. It may be easily cleaned or repaired.
c. Its writing may be shaded from light to heavy.
4. Pen B is inconvenient.
a. It has to be filled frequently.
b. A supply of ink is often not available when it is empty.

III. A balancing of these data shows that
A. For the daily usage of most people Pen A is preferable.
B. For some uses and some people Pen B is preferable.

III. (Same as for Plan 1)

EXERCISES

A. *Make two outlines applying plans I and II to the comparison of the merits of two devices, methods, processes, proposals, etc. with which you are familiar.*

B. *Write one of the two papers outlined. Use headings.*

Notes

1. "Things" is the general term used throughout the remainder of this chapter. Substitute "materials," "machines," "methods," "processes," etc., as the subject for analysis may require.

9

Mechanisms

Descriptions of mechanisms are so common in scientific literature that you may appreciate some help in preparing them. In these instances the objectives in writing are numerous, and the extent and kinds of detail used vary greatly. Some of the kinds that occur frequently are defined and illustrated below.

KINDS OF DESCRIPTION

Descriptions of mechanisms may conveniently be classified according to the author's purpose as *impressionistic* and *scientific.*

In *impressionistic* description the author's purpose is not so much to describe the mechanism objectively as to describe the impression it makes upon someone. Here is an illustration:

> The Eskimo's greatest pride was a telescope, one of those long collapsible marine telescopes made of bronze that looked as if it was the one Nelson had put to his blind eye at the battle of Copenhagen. When it was passed around for our admiration, I saw with pride (Lord forgive me!) that it was of French manufacture.
>
> —Gontran de Poncins, *Kabloona,* p. 217

Note that the author, a Frenchman, emphasizes the attitudes of the Eskimo and himself toward the telescope. That emphasis shows that his purpose is not to describe it in words in the same impersonal way that a camera would take a picture of it. Though there are some objective descriptive words—"long," "collapsible," "made of bronze"—the main purpose is to tell what observers thought and felt about the instrument.

In *scientific* description, on the other hand, the impressions made upon the emotions are disregarded. Words are used in the same way that the mechanical draftsman uses lines, to represent exactly the material aspects of objects. The difference between scientific description and impressionistic description is just the difference between the mechanical drawing and the artist's oil painting: the latter is "art" because it embodies the artist's personal feelings; the mechanical drawing is not art because it is completely impersonal.

In the followng passage there is a blending of impressionistic and scientific description:

> For the first of many, many times I looked at the seal-oil lamp, at the warm gentle glow that rose from the wick floating in the blubber, and there descended into me an affection for this primitive utensil that I am sure will never leave me. This lamp is not a cruet but an open vessel hollowed out of soapstone and filled with seal blubber which melts as the flame heats it. The wick, made of a sort of cotton grown on the tundra, is shaped with the fingers into a saw-toothed length and floats along the edge of the vessel just above the rim. For more light, you lengthen it; for less it is made shorter. When the lamp smokes the wick is of course too long; with a stick made of soapstone you crush it down. If the lamp sputters, the vessel wants replenishing. You put your hand into a barrel of blubber, take out two or three chunks, and let them down into

the vessel where they drop with a soft thud. No one who has not lived with this lamp within the confines of a small space, tent or igloo, can know the radiance it creates, the friendliness and intimacy that radiates from it.

—Gontran de Poncins, *Kabloona*, p. 41

OBJECTIVES IN SCIENTIFIC DESCRIPTIONS OF MECHANISMS

The scientific description of a mechanism may be of two kinds, *generalized* and *specific*.

The *generalized* description is limited to those characteristics peculiar to the genus, the whole class, not to those peculiar to any species or any one member of the class. Thus a generalized description of an airplane would concern its weight, its power plant, its wing surfaces, and the theory of its flight. It is equivalent to an expanded logical definition: An airplane is a "fixed-wing aircraft heavier than air that is driven by a screw propeller or by a high-velocity jet and supported by the dynamic reaction of the air against its wings" (*Webster's Third New International Dictionary*). If the genus were narrowed to, let us say, a *passenger transport plane,* the generalized description would include many more details.

The *specific* description applies to only one species or perhaps to only one member of a species. The complete description of a Chevrolet sedan, 1970 Impala model, would include many details and yet it would fit every one of the thousands of that model sold. That would be a description of a *species;* but a description of the high schooler's jalopy with its special paint job, rebuilt engine, and numerous accessories, even though it was originally a 1970 Chevrolet sedan, would have to be much more particularized.

Your purpose will determine which kind of description you should write. In a textbook on the testing of materials a description of a Brinell hardness-testing machine would probably be limited to its use and the principle of its operation, without details about its construction. But in an application for a patent on a newly invented machine the description must be exact and complete.[1] Adequate drawings must supplement the description.

Your purpose will indicate how detailed the description should be. If your readers need merely to *understand* what an extensometer is, for example, you will include fewer and less exact details than if they wished to construct a duplicate of a certain extensometer.

Your purpose is governed by the needs of your audience. Depending on the readers' needs, the description of a mechanism may vary all the way from a mere sentence definition to the complete and detailed explanation necessary to make a duplicate.

PREREQUISITES FOR WRITING DETAILED SCIENTIFIC DESCRIPTIONS OF MECHANISMS

The prerequisites for writing detailed scientific descriptions of mechanisms are:

1. Clear idea of your purpose, as shaped by reader's needs
2. Intimate knowledge of the mechanism
3. Access to the mechanism while writing
4. Knowledge of terminology
5. Definite plan of procedure

Clear Idea of Your Purpose, as Shaped by Reader's Needs

The needs of your reader should determine many of the characteristics of your detailed description. The structure of your description, the kinds and amounts of details you include or exclude, and the concepts you mention or do not mention, explain or do not explain, should all be matters of choice on your part, based principally on the needs of your reader. Refer to the section on audience in chapter 3 for more detail on the needs of various kinds of readers.

Intimate Knowledge of the Mechanism

Before attempting to write a detailed scientific description of a mechanism, make sure that your knowledge of it is complete. You cannot describe a voltmeter fully if you have never taken one to pieces and examined what is inside. Even then a complete description may be impossible without consulting the manufacturer, for you will require adequate data for each part of the mechanism on all the following points:

Size
Shape
Material
Finish
Connections and relations with other parts
Function

Access to the Mechanism While Writing

Unless you have the mechanism before you while you write, you will be unable to supply complete details of size, shape, material, and finish. Measuring devices such as a ruler, a tape measure, and inside and outside calipers will be almost indispensable.

Knowledge of Terminology

You need to know the names of all the parts in order to avoid writing vaguely of "pieces," "portions," "sections," and "what-you-may-call-thems."

Definite Plan of Procedure

Finally you will need a definite plan of procedure, a clear notion of what to do first, what second, and so on. Probably the worst fault in most descriptions of mechanisms is this lack of plan.

PLAN FOR DETAILED SCIENTIFIC DESCRIPTION OF A MECHANISM

The following plan, though perhaps not the only logical one, is adequate for the description of any mechanism, from a pair of pliers to a power plant. The differences will be in the length and the complexity, not in the arrangement of parts.

I. Introduction
   A. Definition
   B. Generalized view (use of analogy when possible)
   C. List of parts
II. Description of parts
   A. Description of part 1
      1. Definition
      2. General view
      3. List of pieces making up part 1
         a. Description of first piece
            (1) size, shape, material, finish, connections, function
         b, c, etc. Description of other pieces
   B, C, etc. Description of parts 2, 3, etc.
III. Description of the mechanism in use
IV. Conclusion

The following suggestions illustrate the use of this outline, assuming that the mechanism is a simple one which can be described adequately in a few hundred words. For a more complex mechanism all parts of the description would have to be expanded. Use judgment in adapting the outline to your needs—it may be unnecessary to include every topic listed.

Introduction

Begin by defining the mechanism, usually in terms of function. A single sentence may be enough.

Next, without going into detail give your reader a notion of the size, shape, and general appearance of the mechanism. If it looks like something with which the reader is likely to be familiar, use a comparison or analogy. For this part you may need from one to five or six sentences.

Finally, list in a single sentence the parts of the mechanism. The list should be not less than two items, ordinarily not more than five or six. If the mechanism has many parts, you should group the closely related ones as a single item. The order of arrangement in the list should be logical and identical with that in Part II.

Description of Parts

In the body of the paper, take up methodically and in the same order the parts of the mechanism listed at the end of the introduction. In describing each part proceed as if you were describing a separate mechanism: define the part when necessary, describe it as a whole, and finally, if it in turn is made of separate pieces, list and describe the pieces. Whether it is a part made up of more than one piece or a single piece, give the details of size, shape, material, finish, connections, and functions.

If all parts of the mechanism are made of the same material and are finished alike, you can include these details in the introduction and thus avoid unnecessary repetition.

Description of the Mechanism in Use

With all the parts described, your next step is to describe the mechanism in use. The logical method of organization here, where possible, is to follow the path of the force, momentum, or energy transmitted. Show the operation of each part and explain its function in the operation of the mechanism as a whole.

Conclusion

If any conclusion is desirable, it may refer to the uses of the mechanism, its novel features, its cost, its history, or some other significant general aspect.

PLAN FOR A GENERALIZED DESCRIPTION OF A MECHANISM

The plan for a generalized description of a mechanism—one that applies to any member of a class—is similar to that for detailed description. One section (II) should be added so that the main divisions of the outline will be the following:

I. Introduction
II. Theory of operation
III. Description of parts
IV. Description of the mechanism in use
V. Conclusion

The main differences between the two kinds of description are these:

1. A complete explanation of the theory of operation (the general law applied in its use) is needed, probably with a separate main heading.
2. Details of size, shape, material, and finish will necessarily be much less exact.
3. Drawing will be "schematic"—not exact and not drawn to scale, but showing the functions and relationships of parts.

Because of the lack of details in the generalized description, much more complicated mechanisms may be analyzed in the same number of words.

USE OF HEADINGS

Use whatever main heads and subheads your description calls for. No main heading "Introduction" is needed: the reader will understand that a paragraph or paragraphs preceding the first heading are general and introductory. See chapter 4 on the uses and values of headings.

Probably you will need main heads for *explanation of theory, description of parts*, and *mechanism in use,* and certainly subheads under *description of parts.* A final main head *summary* or *conclusion* may be used, or the equivalent in effect can be secured by omitting the heading and leaving quadruple spacing before the last paragraph.

USE OF VISUAL AIDS

You can and should use visual aids—diagrams, pictures, etc.—to supplement the verbal description. You can represent many things more clearly and accurately in a drawing or photograph than in words.

A description that consisted only of drawings and pictures, however, would be inadequate. Of the six qualities to be covered in any exact description of a portion of a mechanism, four—material, finish, connections, and function—cannot be completely indicated by a drawing.

The fact that an application for patent requires a complete *written* description of the device is evidence of the need for verbal details. Section 4888 of the revised statutes of the United States Patent Office reads as follows:

> Before any inventor or discoverer shall receive a patent for his invention or discovery, he shall make application therefor in writing to the Commissioner of Patents and shall file in the Patent Office a written description of the same and of the manner and process of making, constructing, compounding, and using it, in such full, clear, concise, and exact terms as to enable any person skilled in the art or science to which it appertains, or with which it is most nearly connected, to make, construct, compound and use the same. . . .

Engineers are prone to rely too exclusively on drawings and to omit or skimp the written description. The procedure strongly recommended is as follows:

1. Provide mechanical drawings, photographs, perspective drawings, etc. to represent the mechanism as completely as possible. Label or number the parts to facilitate reference to them in the verbal description.
2. Write as complete and accurate a description as possible, covering all details of the six points enumerated—size, shape, etc.
3. Correlate the description with the illustrations by adequate cross references: "as indicated in Figure 1, B," "(Figure 2)," "handle 16 of the plunger 17," etc.

Refer to the section in chapter 4 on the value of visuals for conventions on numbering, labeling, and placement of visuals. The usefulness of your visuals often depends on the extent to which you conform to those conventions.

STYLE IN SCIENTIFIC DESCRIPTIONS OF MECHANISMS

The main objective in writing scientific description is *clearness.* All other qualities of style are secondary. If, for example, repetition of a word or phrase is necessary for clearness, do not hesitate to repeat it, even at the sacrifice of euphony.

At the same time, however, try to make your description pleasant to look at and to read. The following points in particular you need to keep in mind:

1. Vary your sentence structure and length. Begin some of your sentences with modifiers rather than grammatical subjects.
2. Subordinate points of minor interest and importance. Instead of writing

The blast furnace is 100 feet high and 20 feet in diameter. It is tubular in shape. The furnace is lined with firebrick.

subordinate the less important details:

The blast furnace, a tube of steel 100 feet high and 20 feet in diameter, is lined with firebrick.

3. Use the principle of parallel structure when it is possible to coordinate items of equal importance. For example, if you are describing size, shape, and material, see the two samples printed below. The first does not use parallelism, the second does.

The arm is 4½ inches long. It is stamped from a strip of mild steel 0.065 inch thick. The strip is ½ inch wide. Each end is rounded to a ⅝-inch radius.

The arm is 4½ inches long, stamped from a strip of mild steel 0.065 inch thick and ½ inch wide, and rounded at the ends of a ⅝-inch radius.

4. Study the sections on Abbreviations, Hyphenation of Compound Adjectives, and Writing of Numerals in chapter 19. Giving careful attention to such details will result in much better papers.

See chapter 20 for explanations and illustrations of common faults in sentence structure.

EXERCISES

A. *Choose a simple mechanism which you can describe completely in not more than 500 words. It should have at least three parts, one of which moves in relation to the others. Try to make your description so complete that a workman with the proper materials and tools could duplicate the mechanism with only your description as a guide. But do not write a set of directions for making it: use the third person, not the second. In revision, attend carefully to the directions for using visual aids and to the suggestions on styling in this chapter.*

B. *Write a generalized description of a mechanism—one that will fit any member of the species. Choose a mechanism that applies a known scientific law, such as the following:*

hydraulic jack, electric clock, gyroscope, governor, thermostat, surveyor's level, barometer, suction pump, still, hydrometer, rocket

C. *Determine which of the following descriptions is specific and which is general. How does each fall short of fulfilling the directions given in either Exercise A or Exercise B? Have the directions on styling and on the use of visual aids been followed?*

THE CENTRIFUGAL PUMP

The centrifugal pump is a machine that forces a fluid from one place to another by rotating it around a central axis. It is used for many purposes; for example, to draw water from a well, to force water through a fire hose, or to raise water into a tower. It is a spirally shaped shell covering a rotating disk which propels the fluid outward toward the inner end of the nozzle. There are four main parts: (A) the case, (B) the entrance, (C) the impeller, and (D) the nozzle. See Figure 1.

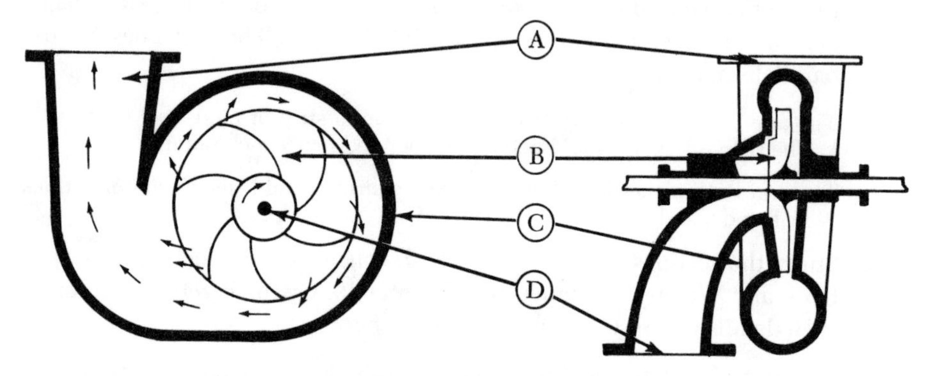

Figure 1. A centrifugal pump.

The Case

The case is the metal shell surrounding the impeller to direct the fluid toward the outlet of the pump. It is usually made of cast iron, but for high pressures may be of cast steel. Its spiral shape directs the fluid with little loss of speed. It is like a snail shell, similar in shape and design. Its size varies with the capacity of the pump. To permit inspection and repair of the impeller, the case is cast in two sections bolted together.

The Entrance

The entrance (B) is connected by a pipe to the source from which liquid is being pumped.

The Impeller

The impeller is a rotating device that forces the fluid outward to the nozzle. It is usually a round disk on which are mounted spirally shaped blades, usually cast in one piece and made of iron, brass, or bronze. Because bronze does not corrode, it is the best material because it is long-lived and retains a smooth finish that lets the fluid flow most efficiently. A rough finish impedes the flow. The impeller has two parts (Figure 2): (A) the disk and (B) the blades.

The Disk. The disk is a round metal plate varying in diameter from a few inches to a foot or more, depending on the size of the whole pump. To it are attached the blades over which the fluid flows. Usually the disk and the blades are cast in one piece, but in pumps used for purposes such as dredging where the impeller is subjected to great wear, the blades are riveted to the disk.

The Blades. The blades are metal devices used to impel the fluid outward from the center of the disk. They are spiral-shaped. Riveted blades, usually made of sheet steel, are less expensive to replace than an entire impeller.

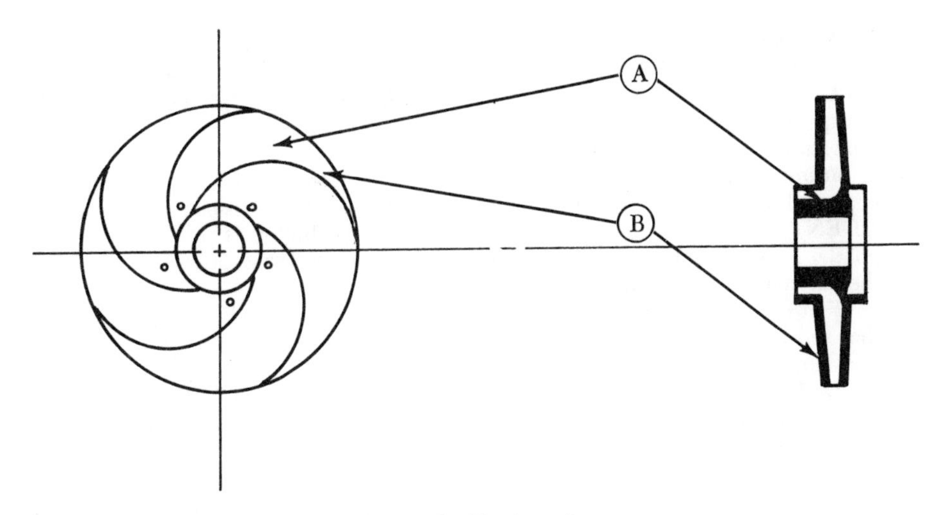

Figure 2. The impeller.

The Nozzle

The nozzle, the opening through which the fluid leaves the pump, is equipped for the connection of a pipe to carry the fluid away at the pressure given it by the impeller. It varies in size with the capacity of the pump.

Operation of the Pump

After entering the pump, the fluid flows over the impeller whose blades force it outward away from the center toward the case. When the fluid strikes the case, it is directed around toward the nozzle by the spiral shape. It leaves the pump at the nozzle at a high speed which enables it to travel great distances under pressure.

Although centrifugal pumps are not new, only in the recent past have they been much improved or widely used. They are high-speed machines, and until the electric motor was invented no motive power has been well suited for them. Since they can lift fluid from a few feet up to several thousand, they fill a wide variety of needs.

A TRANSPONDER

A transponder is an electrical device used to reflect and intensify radar signals. It is most often used by a pilot to make his airplane more "visible" to radar. It is a small device from 6 to 8 inches wide, 3 to 4 inches high, and up to 12 inches long. In appearance it seems to be just a black box.

List of Parts

The transponder (Figure 1) has few parts: an on-off switch, an automatic identification switch, a manual identification button, a frequency dial, and frequency controls.

On-Off Switch. The on-off switch (Figure 1-A) controls the power of the transponder. It is a simple slide-type electrical switch, with the path of travel up and down. For the transponder to operate, the switch must be in the "on" position at the top.

Automatic Identification Switch. The automatic switch (Figure 1-B) controls the reply signal of the transponder. It is a toggle-type electrical switch labeled "RPLY," an abbreviation for "reply." When the switch is in the "on" position, the transponder will automatically reply to a radar signal from the ground. The signal shows up as a blip on the ground radar screen.

Manual Identification Button. The manual identification button (Figure 1-C) is a push-type electrical switch with an automatic spring return. It is labeled "IDENT," an abbreviation for "identify." When the button is pushed, the transponder puts out an intense signal which the radar groundcrew can distinguish from the automatic reply signal.

Frequency Dial. The frequency dial (Figure 1-D) is a cylinder-type indicator that shows on what frequency the transponder will operate. The numbers on the dial show the frequency in megahertz (MHz).

Dial Controls. The dial controls (Figure 1-E) are operated by hand. When the controls are twisted, the frequency dial moves. Different numbers appear, indicating different frequencies of operation.

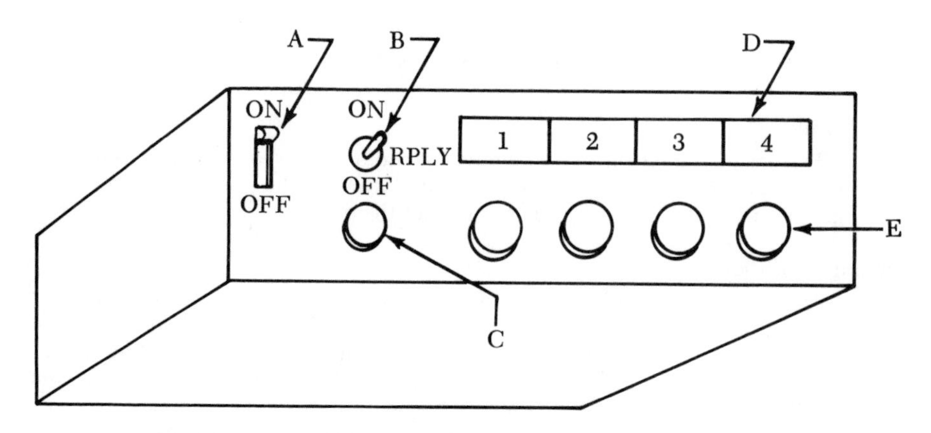

Figure 1. The transponder.

The Transponder in Use

When the transponder is used, the on-off switch is turned on. Unless a strong signal is desired, the RPLY switch is then pushed to the "on" position. If a strong signal is needed, the IDENT button is pushed at intervals. The proper frequency is selected by twisting the frequency controls until the desired frequency is indicated on the dial.

The transponder is helpful in air-traffic control. With it, a controller on the ground is able to see where planes are on the radar screen. Without it, some airplane might escape radar detection. Unseen traffic is a major cause of collisions in the air. Widespread use of transponders makes control of air traffic more effective.

—Larry Priest

A WIRE STRIPPER

A wire stripper, which operates in the same fashion as an ordinary pair of scissors, is a tool to remove insulation from electrical wire (Figure 1). The sharpened notches A and B in the two 5⅛-inch blades cut the insulation around the wire so that it is easily removed.

Description of Parts

The wire stripper has four parts: the main blade, the adjusting blade, the rivet, and the adjusting mechanism.

The Main Blade

The main blade is made of 0.0625-inch sheet steel, ground and cut 5⅛ inches long, as shown in Figure 2. One end F is formed as a 19/64-inch-radius semicircle, the other end G as a 19/64-inch radius quarter-circle. Along the center line and 4¼ inches from the semicircular end, a 3/16-inch hole H is punched to provide the pivot point. The sharpened portion of the blade begins at the corner nearest the

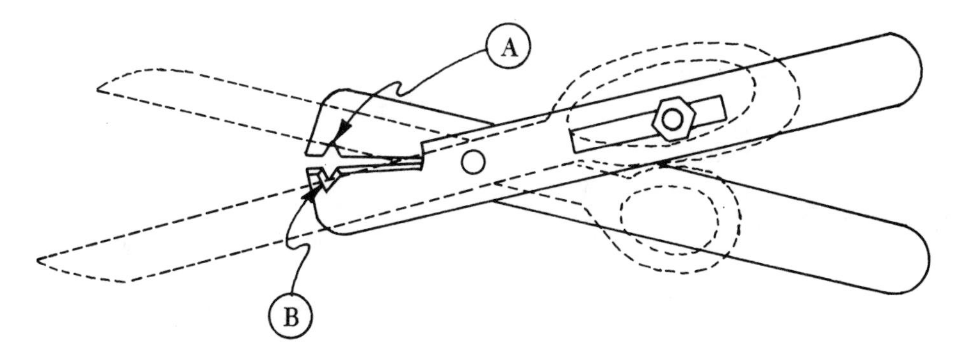

Figure 1. Similar operation of scissors and wire strippers.

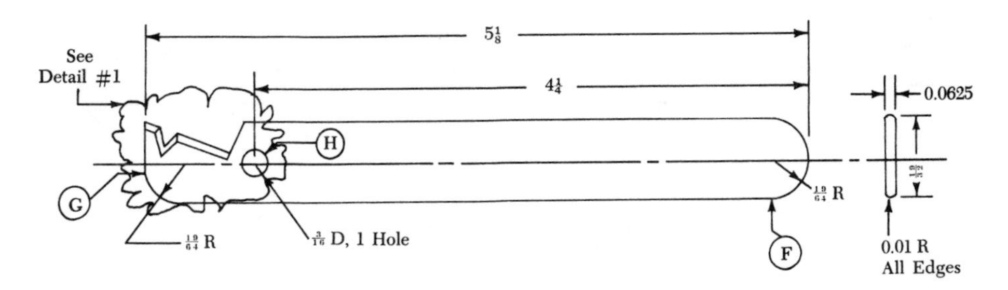

Figure 2. The main blade.

quarter-circle and continues 22/32 inch in direct line with the center of the pivot hole H (Detail 1). A 60-degree notch R is cut in this length 6/32 inch from the corner to a depth of 1/8 inch (Detail 1). The notch and the edges are sharpened to a 45-degree angle. Other edges and corners of the main blade are rounded to a 0.01-inch radius.

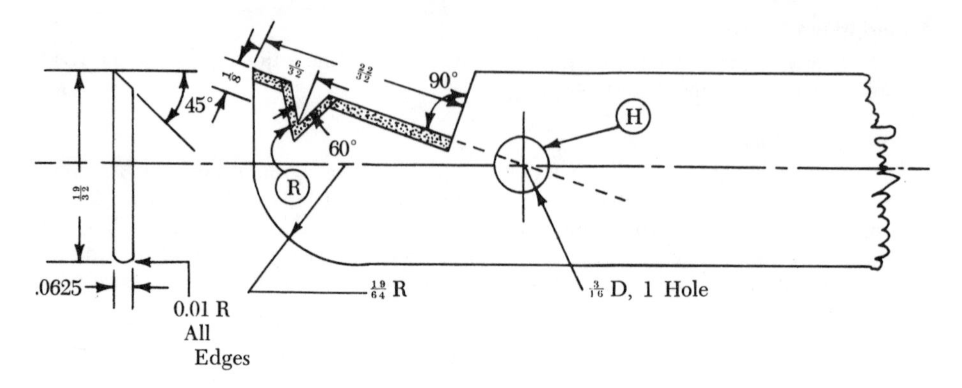

Detail 1. Cutting edge.

The Adjusting Blade

The adjusting blade differs from the main blade only in that it has a rectangular slot N cut symmetrically about the center line 2 1/2 inches from the semicircular end (Figure 3). This 3/16- by 1 1/8-inch slot is for the adjusting mechanism.

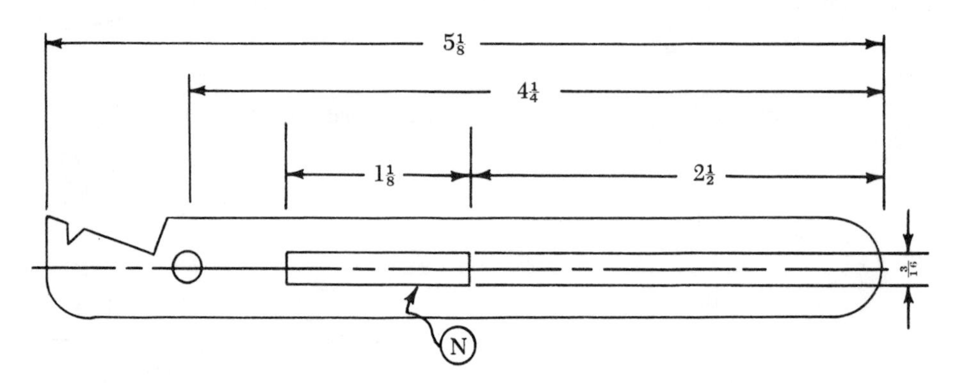

Figure 3. The adjusting blade.

The Rivet

The rivet is a 3/16-inch-diameter by 3/16-inch-long steel buttonhead type (Figure 4).

The Adjusting Mechanism

The adjusting mechanism, placed in the slot shown in Figure 3, consists of a bolt and a burr.

The *bolt* is a standard round-head 8-20 bolt, 5/32 inch long (Figure 5).

The *burr* is a standard hex-head 8-20 burr, 3/32 inch thick and 5/16 inch across the flats, with two notches cut 1/32 inch deep and 1/16 inch in from two opposite sides. This burr fits in the slot cut in the adjusting blade (Figure 6).

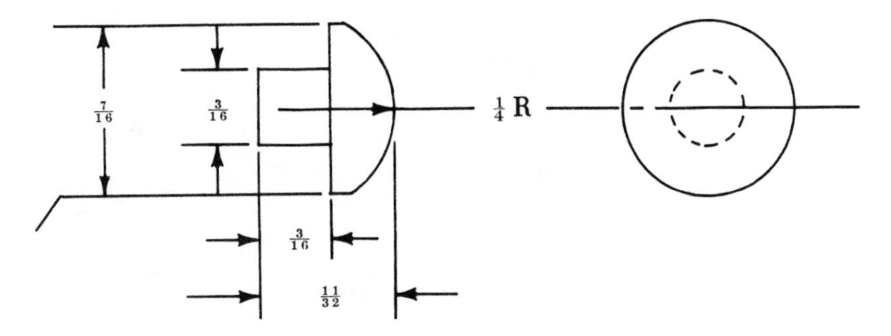

Figure 4. The rivet.

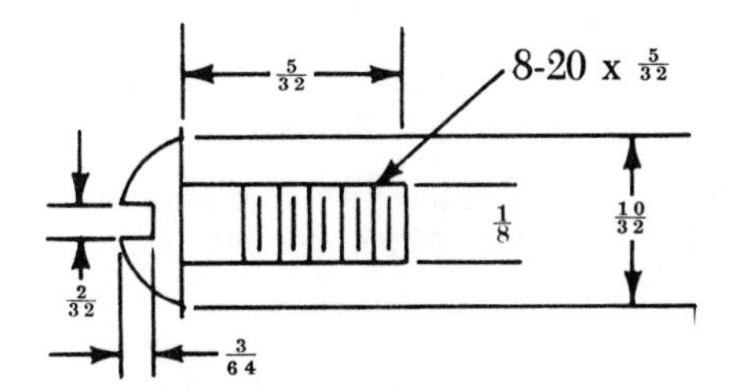

Figure 5. The bolt.

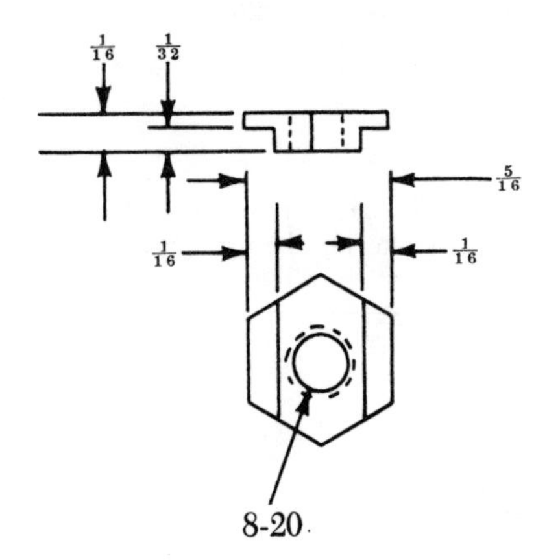

Figure 6. The burr.

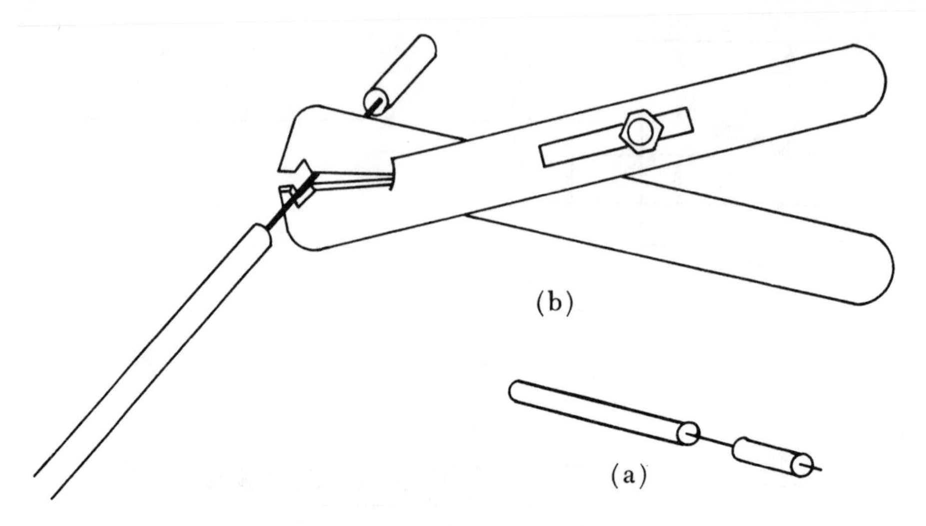

Figure 7. (a) Appearance of the wire after the use of the stripper, (b) Application of the stripper.

Operation of the Stripper

The wire to be stripped is put in the notch, and the stripper is closed upon it. The adjusting mechanism is set so that the sharpened notches cut the insulation but not the wire. When the tool is properly set, its use can result in a cleanly stripped section of insulated wire, as shown in Figure 7.

—R. Hala

Notes

1. Except that details of size and shape are omitted. If these were given, it would be easy to infringe on the patent by altering them slightly.

10

Processes

Most of the general statements about description made in the preceding chapter about mechanisms apply as well to descriptions of processes. The general principles of description are the same, whatever the subject may be.

KINDS OF DESCRIPTIONS

Descriptions of processes may be classified in the same way as descriptions of mechanisms: *impressionistic* and *scientific*. In the first the purpose is partly to give information, partly to describe the impression made upon the observer; in the second the purpose is solely to give information.

An illustration of impressionistic description is the following passage from "Vulcan," a chapter from Joseph Husband's *America at Work*:

> The doors of the hearth were thrown suddenly open. A blinding whiteness streaked with saffron, and heat almost beyond endurance, made me draw back behind a column. A workman thrust a pair of deep-blue glasses in my hand. Slowly the great ladle bent forward. From its spout a trickle of fluid iron poured faster and faster until the white cascade, at full flood, seethed into the hearth-bath. A shower of sparks, strange flowery pyrotechnics, shot high into the gloom. Through the blue glass I peered into the hearth. Like an infernal lake it swirled and eddied, a whirlpool of incandescent flame. Leaping tongues of pink and lavender danced in the blue darkness. Shielding their goggled faces from the heat, the workmen cast lumps of rich ore into the hearth-mouth—black silhouettes of men against the blue glare of an uncanny firelight.

This is good description, but it is not scientific description. You can spot the many words in the passage that report subjective impressions rather than objective fact.

OBJECTIVES IN SCIENTIFIC DESCRIPTION OF PROCESSES

Depending upon the purpose of the writer (again, as shaped by the reader's needs), scientific descriptions of processes may be *generalized* or *specific*. The generalized account summarizes the process, usually rather briefly, the purpose being to define rather than to give the detailed kind of description necessary for complete understanding. The definition of the "Bessemer process," from *Webster's Third New International Dictionary* is of this type:

> A process of making steel from pig iron by burning out carbon and other impurities through the agency of a blast of air that is forced through the molten metal, the blowing usu. being continued until nearly all the carbon is removed, the desired proportions being restored together with manganese by adding ferromanganese while the blown metal is being poured into a large ladle from which the ingot molds are filled.—By permission. From *Webster's Third New International Dictionary,* © 1976 by G. & C. Merriam Co., Publishers of the Merriam-Webster Dictionaries.

Contrasted with the generalized account of a process, the *specific* description has as its objective such completeness that the reader perhaps would not be able to perform the process from the data given but could comprehend every step in detail. If the purpose is to enable the reader to perform the process, the description will probably be put in the form of a *set of directions.*

Here again the author's purpose and the reader's need will govern the amount of detail. Most processes are carried out with the aid of machines or tools: if the purpose is to describe with complete detail, a description of the machine or tool (unless it is very common and description is unnecessary) must be included. Perhaps a description of the process of sawing a board in two with a handsaw would not need to include a description of the handsaw, though an understanding of the shape and action of the teeth would be helpful. But a description of the Rockwell test for hardness would be incomplete for the average reader without a picturization of the instrument with which the process is carried out. The following description of the Rockwell test seems to assume that the reader has access to the machine and would be classed as a *generalized* description.

> The Rockwell test for hardness is a penetration test in which a diamond cone is used for hard materials, and a hardened steel ball 1/16 inch in diameter for soft materials. A minor load of ten kilograms is first applied, which seats the penetrator firmly on the surface of the specimen. The zero of the recording dial on the depth-measuring gage is then set at the pointer and a major load is applied. The major load is 150 kilograms for the cone and 100 kilograms for the steel ball. The major load is allowed to remain until the pointer comes to rest and is then removed . . . .
>
> —Clapp and Clark, *Engineering Materials and Processes,* pp. 21-23

How the load is applied and how the dial functions are not indicated.

The *specific* description of a process, like the specific description of a mechanism, applies to the *species* and not to the *genus.* The process of making pottery (*genus*) has a thousand different variations; even the process of making one particular size and shape of vase (*species*) may be performed in different ways by different potters. The *specific* description would apply only, for example, to Josiah Wedgwood's process as carried out in Burslem, Staffordshire, England, in the late eighteenth century. If the description is to cover all the variations, it is necessarily generalized. Both types of description, of course, are justifiable for different purposes.

PREREQUISITES

The prerequisites for writing detailed scientific description of processes are:

1. Clear understanding of the reader's needs
2. Intimate knowledge of the process
3. Knowledge of the terminology
4. A definite plan of procedure

These are the same prerequisites as in writing detailed descriptions of mechanisms, except that "access to the process" while writing is not included. Only when it involves exact measurements and descriptions of shapes would access to the process be necessary while writing. Intimate knowledge of the process and of the proper terminology to designate its different parts is essential.

PLAN FOR DETAILED DESCRIPTION OF PROCESS

Since the description of a process naturally takes the form of the "story" of the process, the various steps in the order in which they are performed constitute the logical divisions of the outline. Though there is less likelihood of confusion here than in describing a mechanism, you need to pay special attention to the division points. Divide the whole process into "steps," not less than two nor usually more than five or six, and each step, if necessary, into its parts.

- I. Introduction
  - A. Definition of the process
  - B. General information: where, when, by whom performed
  - C. Preparation for the process
    - 1. Materials and apparatus
  - D. Precautions to insure safety, quality, etc.
  - E. List of the steps
- II. Description of the steps
  - A. Description of Step 1
    - 1. Definition
    - 2. General statements about the step as a unit
    - 3. Materials, apparatus, etc. needed
    - 4. List of divisions of Step 1
      - a. Description of first division
      - b, c, etc. Description of other divisions of Step 1 in order
  - B, C, etc. Descriptions of Steps 2, 3, etc.
- III. Synthesis: summary of the steps as a whole
  - A. Appropriate concluding statements if needed

You can use this plan for the description of either simple or complicated processes. In the following discussion of the use of the outline it is assumed that you are describing a fairly simple process in a few hundred words. For more complex processes you would expand the proportions of all parts of the description.

Introduction

At the beginning write a definition of the process; for example, "The process of grinding valves in a gasoline engine consists of refitting the valves to prevent leakage." Usually a single sentence is enough.

Make any general introductory statements necessary for an understanding of the process. Describe here any special requirements of space, time, and personnel.

Include next a description of the materials to be used, what they are and where they come from, and of the tools and apparatus necessary. Since most scientific and engineering processes require the use of tools, machinery, and apparatus which may be unfamiliar to the reader, this part of the introduction may require considerable space. If possible list these needs in tabular form; for example:

The following equipment will be needed:

Bench grinder
Angle gauge
Oil can
Several small drills

Such a list is easier to read and to refer to than an ordinary sentence.

End the introduction with a statement listing steps in the process: at least two, usually not more than five or six. If there are too many steps, it becomes difficult for the reader to keep them separated and in order. If necessary, break down each step in turn into its parts.

The sentence listing the steps usually takes the following form:

> The process of building the synchronizing board may be divided into three steps: (1) building the foundation board, (2) mounting the lamps and switches, and (3) installing the wiring.

In the description of a simple process the whole introduction should ordinarily not require more than two or three paragraphs. Only when the apparatus used is elaborate and unusual will more space be necessary.

Description of Steps

The main part of the description is merely the story of the process, broken down into its logical steps. Take up each of the steps in chronological order and define and describe it as a whole, noting the materials, apparatus, etc. for each step. If necessary, break down each step into its parts and take them up in the same way.

Be sure to explain fully the reasons for every step. Since you want the reader to understand the process completely, you should explain why each step is performed as indicated.

Synthesis

After describing the separate steps in turn, briefly summarize the process so that the reader can see it as a whole and not as a series of disjointed parts.

In conclusion make whatever general statements seem necessary or appropriate. Statements about the values or advantages of the process, its difficulties, the alternative ways of performing it, etc. are possibilities.

POINT OF VIEW

In the description of a process you need to adopt and carry through consistently a single point of view. The possible points of view are these:

- I. First person ("I" or "we," autobiographic narrative)
- II. Second person ("you" or imperative, used in giving directions)
- III. Third person
  - A. Active voice ("he," "she," or "they" do so and so)
  - B. Passive voice (things are done by him, her, or them)

First Person

Use the first person in informal, impressionistic accounts of processes, where the person performing the process is important, as well as the actions being performed. The following example is from a sketch entitled "Molten Steel," by C. J. Freund:

> I raised the lever . . . For an instant I was bewildered but immediately realized that instead of merely easing the stream I had shut it off entirely. Al sprang for the ladle but, in my eagerness to correct my error, I bore down on the lever too heavily so that the full stream of metal plunged into the gate at the very moment when the mould could least stand the shock.

In popularized descriptions of processes, where the purpose is to catch the attention and interest by dramatizing the action, this point of view is desirable. In scientific descriptions it is rarely used, however, because the intention there is to focus attention upon the *action* and not upon the *person.* Only when some special historic interest attaches to the performer is the first person likely to be employed, as for example, when William Harvey described experiments in demonstrating the circulation of the blood.

Second Person (Set of Directions)

Use the second person, the imperative, only when giving directions. If you are telling a person how to perform a process, not merely describing the process for comprehension, you properly say, for example, "Drill a ½-inch hole beside each terminal of the single-throw switch" instead of "A ½-inch hole is drilled," etc.

By using this point of view you can write any specific description of process as a set of directions. You can use exactly the same outline for both. See chapter 10 for detailed instructions and illustrations.

Third Person

Active Voice

The third person active point of view, though not much used in scientific descriptions of processes, may often be better than the common third person passive. The active voice permits a degree of human interest and avoids the uncertain agent implied in the passive voice, who, though unseen and unheard, does all and knows all. Note the vigor of the following account of the process of riveting, as reported in *The New Yorker* for August 15, 1942, from an interview with Raymond Davidson, champion riveter:

> The heaters, they heat the rivets in the fire there on the deck. Twenty at a time—every time they take one out they put another in, like perpetual motion. One of them picks a rivet up with his tongs and slings it up to a passer-boy, and he catches it in his bucket. Then another passer-boy takes the rivet with his tongs and gives it to the holder-on, and the holder-on knocks with his machine from inside the hull, for a signal, because I'm outside and of course I can't see him. Then he puts the rivet through the hole and you take your machine—eighteen pounds, it weighs—and drive the red hot rivet in, with all your strength, until it's flush against the plate. Then you lay down your machine and pick up your chipper, and clip off the rivet where it sticks out any, then you take your riveter again and finish her off—flat, smooth, round as a silver dollar. Then the holder-on knocks and out comes another rivet.

The account is all in the active voice, though in different persons. If it were transposed into the passive, "The rivets are heated there on the deck by the heaters," etc., it would lose in color and vigor.

Passive Voice

The third person passive is most commonly used in scientific prose, mainly because it emphasizes the action and the thing done and minimizes the importance of the doer. Since science is concerned primarily with the objective and the impersonal, the passive point of view, though devoid of human interest and color, is ordinarily proper for accounts of scientific processes.

Compare the effects of the two points of view in the following passages, the first in the active voice, the second in the passive voice:

> The welder scratches the electrode rapidly across the surface of the work. As soon as he hears a crackling sound, he lifts the electrode away from the surface. At this instant he begins the bead. To insure a good weld he keeps the electrode from 1/8 to 1/4 inch from the work. He keeps a pool of molten metal in front of the electrode at all times.

> The electrode is rapidly scratched across the surface of the work. When a crackling sound is heard, the electrode is lifted away from the surface. At this instant the bead is begun. To insure a good weld the electrode is kept from 1/8 to 1/4 inch from the work. A pool of molten metal is kept in front of the electrode at all times.

In such a description, where the "person doing" is at least as important as the thing done, the third person active is certainly preferable. Also the active voice usually requires fewer words, and is thus easier for inexperienced writers to write clearly, as well as quicker for busy readers to read.

In the passage following it would be impracticable to substitute the third person active: who did the repairing and the splicing is of relatively little significance.

> The engine main generator and the other power plant and cab accessories were repaired at the Electro-Motive Corporation's plant, and the apparatus was reassembled in the railway's Aurora shops. The entire new front of the power car was spliced to the salvaged and repaired rear end of the original car by spot welding, using lap reinforcement pieces on stress-bearing members.

To avoid unnecessary shifts from active to passive, or passive to active, check the first draft of the paper carefully for consistency.

TENSE

The description of a process will usually be in the present or the past tense: present if the description is general and without reference to a specific past time; past if it refers to a specific performance of the operation in the past. In a set of directions for doing something in the future, you may use the future tense—"you will do so and so"—although the present tense is usually preferable—"do so and so." A shift from one tense to another would be justified only in unusual circumstances.

TELEGRAPHIC STYLE

Do not omit the "little" words, as you would in writing a telegram. Instead of "Turn key in switch," write "Turn the key in the switch."

USE OF HEADINGS

Use headings to indicate the various steps in the process, subheads if the paper is long. If the part of the introduction dealing with "preparation for the process" is long enough, a heading would be desirable for it. The uses and values of such headings have already been enumerated in chapter 4.

USE OF ILLUSTRATIONS

As in the description of a mechanism, illustrations are useful and often necessary in the description of a process. The same procedure in the use of illustrations is recommended for both types of description.

SCIENTIFIC STYLE—REMEMBER THE READER

Remember the reader. Any bit of advice on style is subordinate to that principle. Every word you choose, every sentence you write, should be tested by the question "Will it communicate my meaning?" Decide whom you are writing to: a college freshman, a classmate, a professor in your field, a mathematician, an economist, a chemist, a food technologist, a farmer, etc. Keep your reader constantly in mind.

The rules for variety, economy, and correctness of sentence structure are the same in all types of writing, including scientific. See the discussions in chapter 9, Mechanisms, and in chapter 19, Mechanical Aspects of Style. Certain aspects of style, however, are more important in scientific writing because of the nature of the subject matter and because of the nature of the readers. The subject matter tempts writers to ever-greater complexity, while the readers—who are often reading in haste—demand greater simplicity. These conflicting forces—the one demanding complexity and the other demanding simplicity—make clarity and economy of style particularly important. You should always be sure that at one stage in your revision process you ask this question about each sentence: Is there another way to say the same thing shorter and clearer?

EXERCISES

A. *Write a description of a simple process with which you are thoroughly familiar. Use either the third person active or the third person passive point of view, whichever your instructor specifies.*

B. *Study the organization of material, the use of headings, the point of view, and the correlation between description and drawings in the following analyses. Do you find any incompleteness or lack of clarity in them?*

OFFSETTING PROCEDURE

The continuation of an original line of survey is often prevented by an obstacle of some sort, such as the boulder in Figure 1. The distance A to B must then be determined by a procedure termed offsetting, adopted when a distance is not directly measurable because something is in the direct line of sight. The procedure is often used in surveying and in forest-inventory work.

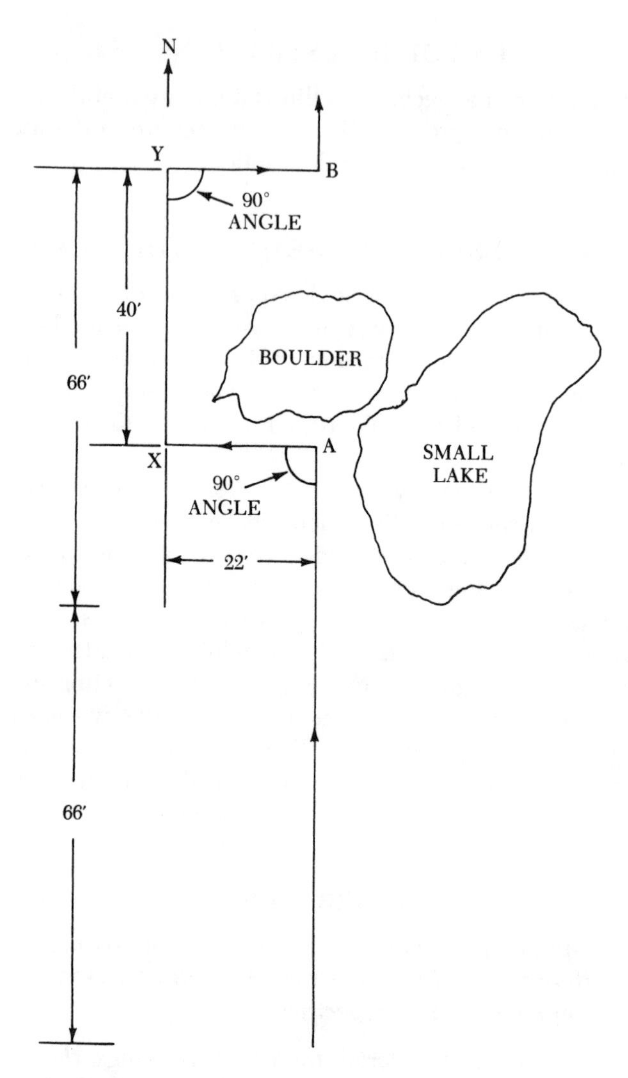

Figure 1. The offsetting procedure.

The only equipment needed is a measuring tape, a compass, and a notebook and pencil.

The offsetting procedure is divided into four steps: (1) extending the survey line to the obstacle; (2) laying off an angle of 90 degrees to either right or left; (3) measuring to a fixed point and extending the survey line from this point; and (4) laying off another 90-degree angle and tying into the original survey line.

Extending the Survey Line to the Obstacle

The survey line is extended to a point A (Figure 1) with little difficulty since there are no obstacles. The distances indicated on the diagram are merely for illustration. The important one to be determined is that from A to B. Usually point A is established far enough from the obstacle (here a boulder) to provide working room when the 90-degree angle is laid off.

Laying Off a 90-Degree Angle

The 90-degree angle is laid off to establish a line perpendicular to the survey line. The perpendicular line can extend to either the right or the left. In Figure 1 it is to the left (west) because another obstacle, the lake, prevents offsetting easily to the right. The angle of offset is 90 degrees only because distances are easier to determine and to measure when the angle is a right angle.

Measuring to a Fixed Point and Extending the Survey

The line from point A is extended to X, at whatever distance from A is necessary to make it possible to extend the line of survey past the obstacle. In Figure 1 the distance is 22 feet, though it might have been 20 or 24. The point X must be established so that the line from X to Y is unobstructed. That line is then extended parallel to the original line of survey and far enough past the boulder to permit an unobstructed view back to the original line. Point Y is arbitrarily established 40 feet from point X.

Laying Off a Second 90-Degree Angle and Tying into the Original Line

At point Y another 90-degree angle is established so that line YB extends to the original line of survey and is parallel to line AX. Point B is then marked 22 feet from point Y, and there another 90-degree angle is laid off.

If the measurements of angles and distances have been correct, the original line of survey can now be extended indefinitely beyond B.

The procedure described provides a simple foolproof way for bypassing an obstacle in a survey line. The accuracy will not be affected by differences in altitude between points in the offset.

—G. Firch

THE LYTIC CYCLE

The lytic cycle is the process of reproduction in the virus, one of the simplest forms of life known to exist. Its genetic material consists of a single strand of nucleic acid contained within the viral headpiece. The tail of the virus serves as a sheath through which the nucleic acid can be transferred into a bacterial cell (Figure 1). All viruses are parasites of bacteria, and viral reproduction can take place only within the bacterial cell that serves as the host. All of the materials used in the process of reproduction are obtained from the contents of the bacterial cell.

The lytic cycle can be divided into four steps: (1) attachment of the virus, (2) entry of viral nucleic acid, (3) multiplication of viruses, and (4) release of viruses.

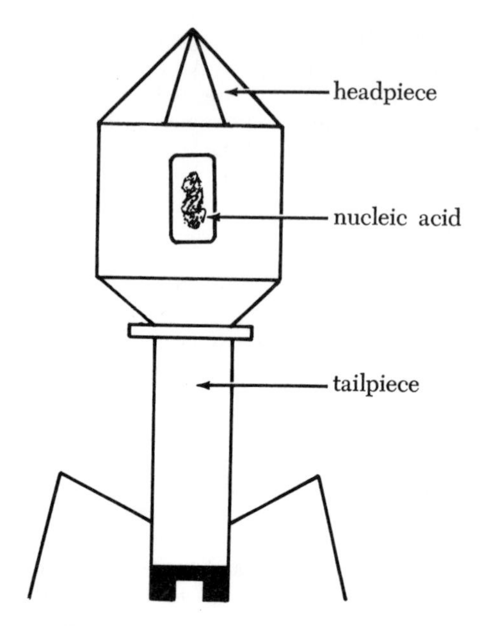

Figure 1. The structure of a virus.

Attachment of the Virus

In the first step, the virus adheres to the bacterial cell. The chemical composition of the cell wall of the bacterium determines the shape of the structures on the wall. These structures serve as reception sites for the virus. If the shape of the viral tailpiece fits the structure of the bacterial wall, the virus becomes irreversibly attached (Figure 2-A). Within a few seconds the second step begins.

Entry of Viral Nucleic Acid

In the second step, the viral nucleic acid enters through the tailpiece (Figure 2-B). The virus releases an enzyme that makes a hole in the bacterial cell wall. The nucleic acid of the virus is then released, and by way of the tailpiece, passes through the opening and into the bacterium. The tail is then retracted. The entry of the viral nucleic acid causes formation of an enzyme that prevents entry of any other virus of that type, and thus renders the bacterium immune to further infection by a virus of the same type. The synthesis of the bacterial enzymes is suppressed, and the enzymes already present are made to serve the needs of the virus. The orifice of entry of the viral nucleic acid now becomes sealed, and the nucleic acid of the bacterium is broken down.

Multiplication of Viruses

In the third step, the virus begins to make use of its host cell for multiplication of viral nucleic acid, headpieces, and tails. These components first appear as separate prefabricated parts (Figure 2-C). During a brief period called maturation, these parts are assembled, and dozens of new, complete, mature viruses are formed (Figure 2-D).

Release of Viruses

In the fourth and last step, after a certain number of viruses are formed, they are released from the cell. The number formed before release is different for each type of virus. When this number is reached, the viral nucleic acid causes the bacterium to produce an enzyme that destroys the bacterial cell wall. The wall ruptures, and the infective viruses are released (Figure 2-E). The lytic cycle ends as the initial virus and the bacterium are destroyed.

Summary

Within a few minutes, the steps of attachment, nucleic acid entry, multiplication, and release are completed, and one of life's simplest forms of reproduction has taken place. The parasitic virus has succeeded in propagating its species, but it has done so at the expense of itself and its host, the bacterium.

—Diane Doubet

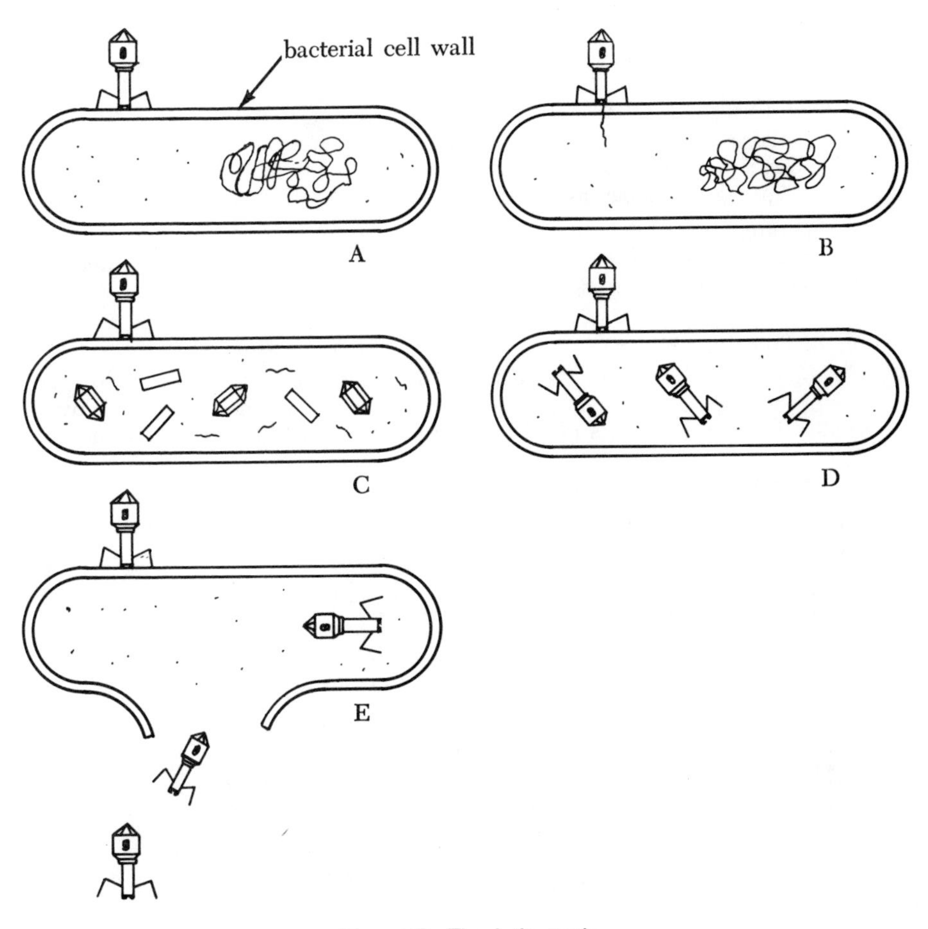

Figure 2. The lytic cycle.

PYROLYSIS OF MINERAL WASTES

Pyrolysis is a chemical process in which various materials are heated to a high temperature in the absence of oxygen and broken down into simpler chemical compounds. These simpler compounds are valuable in themselves and also as raw materials in other chemical processes; thus, pyrolysis is similar to any chemical process that turns out a salable product. It may be applied only to specialized material, such as rubber tires and paper pulp, or to a mixture of materials, such as municipal solid wastes. The last type of application is described below.

Equipment for Pyrolysis

The equipment needed for a pyrolysis waste-treatment plan consists of the following:

Apparatus for pre-pyrolytic treatment

- Metal separator
- Glass separator

Pyrolyzer (Lantz converter)

Apparatus for post-pyrolysis treatment

- Solids separator
- Gas-liquid separator (condenser)
- Liquid-liquid separator

Apparatus for further treatment

Figure 1 below represents an overall view of the process. The four steps are described in the following sections.

Steps in the Process

Pre-pyrolysis Treatment. Pretreatment of municipal wastes is necessary since some of the materials commonly found in a city's refuse are not suitable for pyrolysis. Such materials include metal and glass, both of which form a viscous fluid

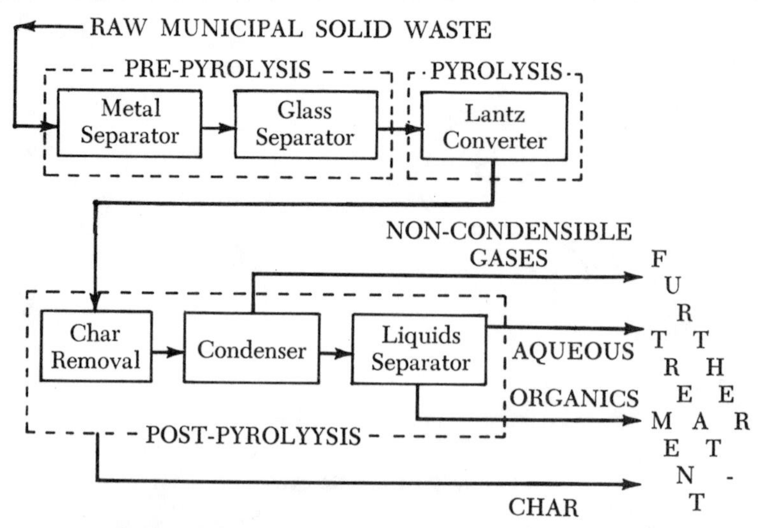

Figure 1. Process flowsheet for pyrolysis of municipal solid wastes.

at high temperatures that would clog the pyrolyzer. Metals and glass, therefore, are first removed from the raw material. Methods to effect this separation might include one or more of the following: sorting by hand, magnetic separation, flotation (glass and metal would tend to sink), vibration tables, and inertial separation. The last procedure operates on the principle that the heavier an object is, the sooner it will fall if thrown into the air.

Pyrolysis. The key element of a pyrolysis plant is the pyrolyzer itself. The prototype for this piece of equipment is the Lantz converter, patented by C. D. Lantz in 1961. It consists of a sealed airtight retort[1] cylinder, about 20 feet long, with an insulated outer shell. Attached to this is the preheating stage, which contains a telescoping feed stage through which the raw waste is introduced. The retort is rotated slowly as it operates, and a temperature of approximaely 1200 F is maintained. The products formed in the pyrolyzer include steam, volatile gases and liquids, carbon dioxide, and charcoal.

Post-Pyrolysis Treatment. The pyrolysis products leave the converter as a mixture of solids, liquids, and gases which must be separated in order to be further utilizable. This separation is effected in three steps. (1) First, the gas and liquids are separated from the solids in a simple settling tank; the solid materials sink to the bottom, and the gas/liquid is drawn off at the top. (2) Next, the hot gas/liquid is run through a condenser, where the condensable vapors join the liquid stream. The non-condensable gases, e.g., carbon dioxide and carbon monoxide, are vented from the top of the condenser. (3) Lastly, the liquid stream is separated into an aqueous stream and an organic-chemicals stream. These two do not usually mix with each other, just as oil does not mix with water. The separation of the converter product into four streams is now complete.

Further Treatment. To realize the economic benefits of pyrolysis, the four streams must be further treated to make salable products. The additional equipment required depends on what products are wanted. For example, the organic stream contains basic alcohols which could be recovered by distillation. The gas stream contains carbon dioxide which could be stripped out with the proper solvent. What is left in the gas stream makes a good fuel gas, similar to natural gas, and could be liquefied and sold. These are not the only possibilities of further treatment.

Summary

The process of pyrolysis as applied to municipal wastes can be summarized briefly in four steps:

1. Removal of material unsuitable for pyrolysis.
2. Heating of the remaining wastes in a Lantz converter.
3. Separation of the converter product into usable streams.
4. Treatment to obtain salable, profit-making products.

The success of the entire process depends mainly on the fourth step. It is probably the most important.

Notes

1. *Retort* is the general term for vessels in which heat is used to cause decomposition of a material.

—David Keeley

11

Sets of Directions

Sets of directions are constantly needed in almost every field. The manufacturer has to prepare them for customers—directions for assembling a product, for using it efficiently, for repairing it when it breaks down. Managers and foremen need them for employees new to their jobs, for persons working on assembly lines, for inspectors of finished products. Teachers prepare them for the convenience of students—directions for doing laboratory experiments, for using instruments, for preparing and writing reports, term papers, and theses. The uses for sets of directions are almost unlimited. Yet a really good set of directions is hard to find.

COMMON FAULTS OF SETS OF DIRECTIONS

A common complaint of customers and employees and students is that they were given no directions, or at best misleading or inadequate directions for their needs. The commonest of these faults are that the writer:

1. Assumes that readers know much more about the subject than they really do.
2. Uses technical terms that are meaningless to the layman.
3. Doesn't separate, number, and paragraph directions so that the reader can follow them easily.
4. Provides no drawings or pictures to supplement the directions.
5. Doesn't place drawings or pictures where they are immediately useful to readers.
6. Fails to refer to drawings when readers need visual aids.
7. Fails to give reasons for the procedure specified.
8. Omits references to general principles that might explain a direction.
9. Doesn't warn readers when there is a chance of making a mistake.
10. Uses a telegraphic style often difficult to understand.
11. Shifts from second person to third without reason.
12. Uses transitional words or phrases that aren't needed.

Thousands of hours are wasted daily while people try to follow poorly written directions. White-collar workers often aren't qualified to write good directions when highly technical machinery and procedures are involved. Engineers aren't interested in the task, or assume that any sensible person should be able to read a blueprint or a diagram, and needs no directions. But someone has to write them, if it is only to explain how to open a can of sardines. As Ken Macrorie said in an amusing essay on "The World's Best Direction Writer" (*College English,* February 1952) ". . . it is an honorable occupation in a dirty business world," but unfortunately it is not a highly remunerative one.

REQUIREMENTS FOR A GOOD SET OF DIRECTIONS

We can infer the requirements for a good set of directions from the preceding section.

1. Completeness

Keep your readers constantly in mind. Are they laymen, executives, experts, or technicians? Whatever the case, write for those within the group who know least about your subject. They will need every aid you can give them. Experts can ignore the directions if they don't need them.

2. Explanation of Technical Terms

Lest any reader may not understand them, define and explain your technical terms. Experts can skip the explanations, but non-experts may be frustrated without them. Most people don't have technical dictionaries to which they can refer.

3. Proper Emphasis

Number each direction and devote a whole paragraph to it, even if it is only a single sentence. Readers can then check them off, thinking, "All right, I've taken that step. Now what's next?" The next numbered paragraph will tell. If you include two or more directions in a single paragraph or under a single number, readers are likely to overlook one of them. Allow plenty of space for each direction.

4. Use of Visual Aids

Include any diagrams, pictures, graphs, etc. that will help readers to follow a direction. (Refer to the section in chapter 4 on Tabular, Graphic, and Pictorial Methods of Presentation.)

5. Reference to Visual Aids

Put visual aids where the reader can most easily refer to them. If such an aid can be on the same page and near the direction it concerns, it will be most conveniently placed. If it is on a different page, readers may not refer to it when they should, and consequently may misunderstand a direction. Your own experience will tell you how irritating it is when a textbook prints a diagram, picture, table, graph, etc. on a different page from the explanation of it. If it is impossible to place it on the same page, put it on the next page.

6. Explanation of Visual Aids

Don't fail to allude to, if necessary explain, any visual aid. Usually it is not enough merely to refer to it, as in writing "See Figure 1." Add any explanation that will help readers to follow your directions. Too often the writer assumes that the meaning of a visual aid is self-evident and leaves many readers to puzzle out its meaning for themselves.

7. Justification for a Direction

Be sure to give the reason for any direction, especially if your readers might think, "Why do that?" Otherwise, they might follow a different procedure which seems at the moment to be preferable.

8. Explanation of General Principles

Explain any general principles that readers ought to understand if they are to follow your directions easily. If the principles underlie the whole procedure, explain them in the introduction. If they refer to only one direction, put the explanation after the direction.

9. Suggestions for Avoiding Mistakes

Warn readers when there is a possibility of making a mistake. Novices especially may need warning to prevent wasting time or ruining expensive equipment. Mistakes due to lack of experience or skill may not be avoidable, but those due to ignorance or carelessness are inexcusable. If, for example, your instruction emphasizes the importance of gluing a wood joint instead of nailing it, you may prevent making something that breaks under stress. Such warnings should be placed at the *beginning* of the set of directions. Do not assume your reader will read all the directions before starting the process. Your own experience tells you the natural tendency to follow instructions step-by-step without first reading over all of them.

10. Avoidance of Telegraphic Style

Don't leave out all words like *the* and *a,* or the subjects or objects of verbs, words necessary for full grammatical expression. Telegraphic style may be justified if space is limited, as on bottles and other small containers. But in the average set of directions you will save less than 5 percent of your space in using it. Furthermore, telegraphic style is likely to obscure the thought. Instead of "Put curd into bowl, add water," it is better to write, "Put the dried curds into a small mixing bowl and add 4 tablespoons of cold water."

11. Consistency in Point of View

Use the second person imperative whenever you are writing a direction. For example, you should say,

> Pour two cups of skim milk into the saucepan.

not

> You should pour two cups of skim milk into the saucepan.

or

> Two cups of skim milk are then poured into the saucepan.

The last two are not imperative constructions.

But if you are stating a general principle or giving some useful supplementary explanation, use the third person. For example, after a direction "Add 6 tablespoons of vinegar to the milk," in the same paragraph could follow the explanation, "The vinegar will sour the milk and precipitate the casein out of the mixture as curds."

12. Use of Clear Transitions

Make all your transitions from section to section and from direction to direction as clear as possible. Proper use of headings, numbers, and paragraph indentations will usually be sufficient. Don't insert unnecessary transitional words and phrases. If direction 3 is "Add one teaspoon of bicarbonate of soda," direction 4 should begin, not with "then" or "now" or "after you have done this," but simply with a new paragraph, the number 4, and the verb in the imperative: "4. Continue the blending, etc." Sequence in time ordinarily governs the arrangement of directions. Only if two or more things are to be done at the same time would a transitional phrase like "at the same time" or some explanation be needed.

PLAN FOR A SET OF DIRECTIONS

The plan for a set of directions is essentially the same as for a description of a process. In developing the plan, however, you need to pay special attention to the division points, both major and minor. Remember that your readers are, presumably, going to perform the process with your directions as the only guide. If they misunderstand a point in a description of a mechanism, perhaps the loss is not serious; but if they misunderstand a direction, they may lose a lot—including their temper.

I. Introduction
   A. Definition of the process
   B. General information: where, when, by whom performed
   C. Preparation
      1. Time, place, conditions
      2. Materials and apparatus
   D. Precautions to insure safety, quality, etc.
   E. List of the steps
II. Description of the steps (a heading for each)
   A. Description of Step 1
      1. Definition and purpose
      2. Explanation of the step as a whole
      3. Materials and apparatus required
      4. Directions (numbered and separately paragraphed)
   B, C, etc. Description of Steps 2, 3, etc.
III. Synthesis: summary of the whole procedure (if needed)
   A. Appropriate concluding statements

The outline is based on the assumption that you are giving directions for a simple process not requiring special skills or complicated machines, and requiring only a few hundred words. For more difficult subjects, like overhauling an automobile engine or building a house, although the same basic plan would serve, hundreds of directions listed under numerous headings and subheadings would be needed.

Since the plans for a description of a process and for a set of directions are similar, no further analysis is included here. See the suggestions for following the outline in the preceding chapter.

EXERCISES

A. *Write a set of directions for performing a simple process. It should be one with which you are thoroughly familiar from having performed it yourself. Use the second person imperative in directions, but the third person in introductory and concluding statements and in explanations.*

B. *Study the organization, use of headings and numbers, transitions, and use of visual aids in the following papers. Do they fulfill all of the requirements for a good set of directions?*

WEIGHING IN GRAVIMETRIC ANALYSIS

Gravimetric determinations are made frequently in analytical chemistry, and require great accuracy. If a precipitate is to be weighed and a quantitative analysis made from the weight, it is obvious that the precipitate must be dry. Not only may water be absorbed on the surface, but it may be trapped in the crystals of the precipitate.

Proper drying vessels, drying temperatures, and procedures are essential to assure accurate, reproducible results. And only repeatable accurate results will lead to meaningful research.

Equipment

The electric muffle furnace is used to dry precipitates that may be easily reduced in currents of air or decomposed at high temperatures. Use of the furnace is the most accurate technique; however, it is also the slowest. The precipitate should be placed in a porcelain crucible in the furnace.

The Meker burner, at its usual burning temperature of 800 C, is a quick, usually available source of heat to eliminate water from precipitates that are not easily reduced or volatilized.

Porcelain crucibles, porous-bottom crucibles, and platinum crucibles are available; but do not use platinum ones for easily reduced precipitates nor for active work since their mechanical strength is small.

Procedure

Break the whole procedure down into the following steps:

First Steps

1. If using the usual porcelain crucible, position it above the burner flame at an inclination, as shown in Figure 1.
2. If using porous-bottom crucibles, place them in a regular crucible as in Figure 2 to minimize the flow of gas through them.
3. Warm the porous-bottom crucibles slowly to remove the water from the pores.

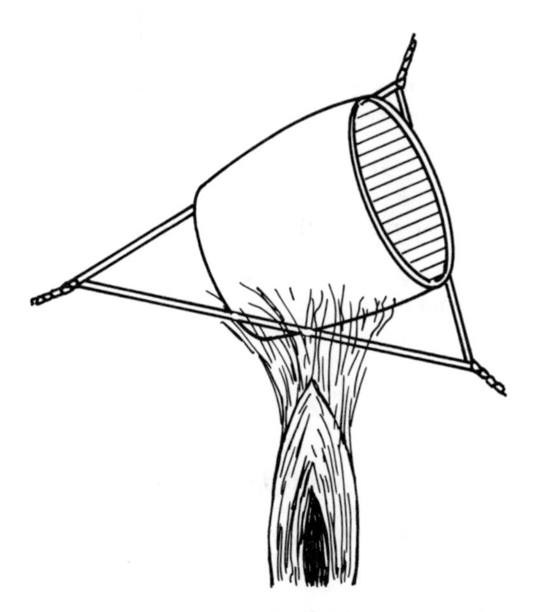

Figure 1. The crucible, tilted and placed at the top of the flame.

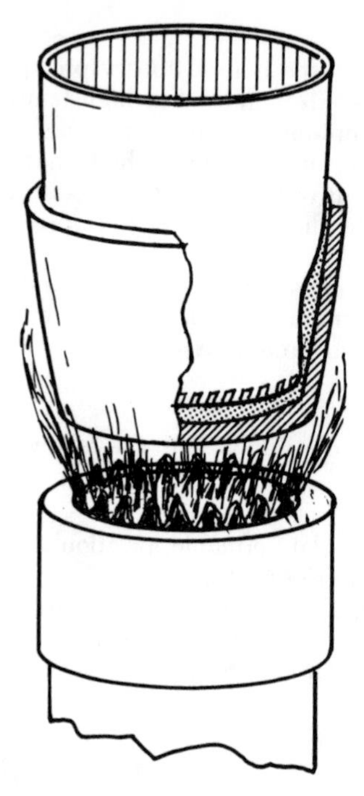

Figure 2. Porous-bottom crucibles, inserted in a regular crucible for heating.

Second Steps

4. When the precipitate reaches the desired temperature, cool the crucible until it can be touched.
5. Put it in a dessicator to cool to room temperature.
6. Weigh the precipitate.

Final Steps

7. Repeat steps 4, 5, and 6 with all samples.

Final Checking

If the weighings made in step 6 do not agree to within 0.5 mg, repeat the procedure. If the results are inconsistent, investigate the possibilities of reduction, volatilization, and decomposition in your procedure.

—Duane J. Erdman

PURGING THE PLASTIC INJECTION MOLDING MACHINE

Periodically you should purge (clean out) the barrel and screw of the plastic injection molding machine to remove small pieces of the plastic molding compound that occasionally stick to the barrel and screw. If these pieces are allowed to remain in the barrel, they occasionally break loose and show up in the molded part, and are responsible for broken corners, odd-shaped holes, and other unacceptable flaws.

You should purge the machine at least once every 4 hours during molding operations. In addition, since these pieces often become stuck to the barrel and the screw when the machine is not running, you should purge it after each break in operation. Whenever these pieces begin to show up in the molded parts, you should purge the machine.

PROCEDURE

The whole procedure can be completed in three steps, preparing for purging, purging, and resetting.

Preparing for Purging

At the end of a molding cycle, after the mold is open, proceed as follows:

1. Raise the safety gate (Figure 1), remove the part, place it in a suitable container such as a box or bag, and leave the gate open. When the gate is open, a bar swings over to prevent the upper half of the mold from coming down (Figure 1).
2. After making sure the safety bar is in place, move to the right side of the machine and switch the controls from automatic to manual (Figures 2 and 3).
3. While still standing on the right side, push the slide valve below the plastic hopper to the right. This shuts off the feeding of molding compound into the barrel and screw (Figure 3).

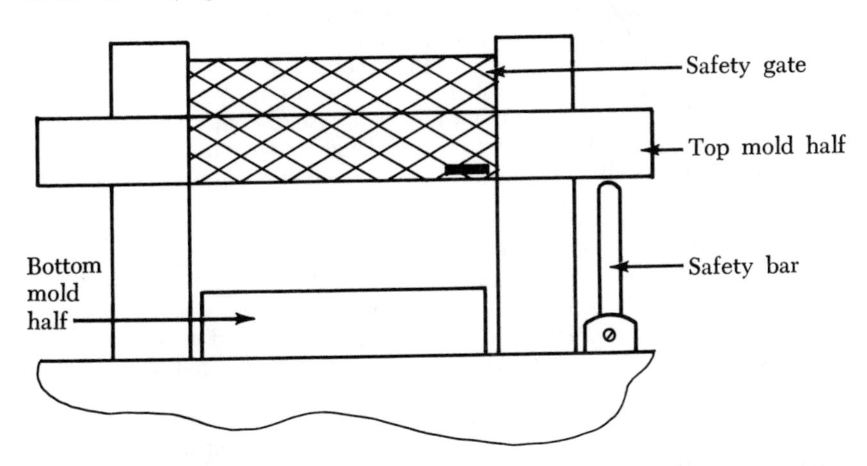

Figure 1. Partial front view of plastic molding machine.

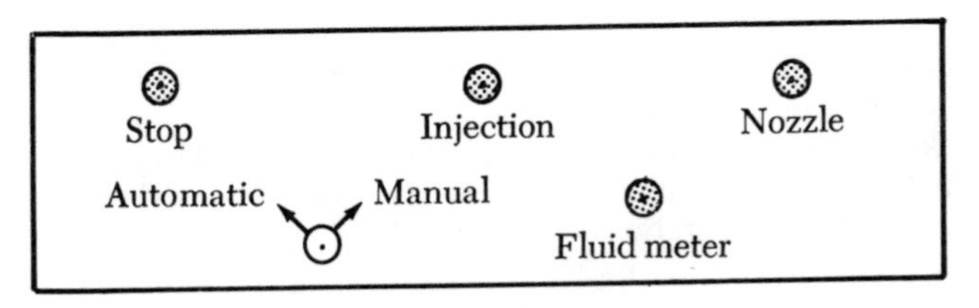

Figure 2. Control panel.

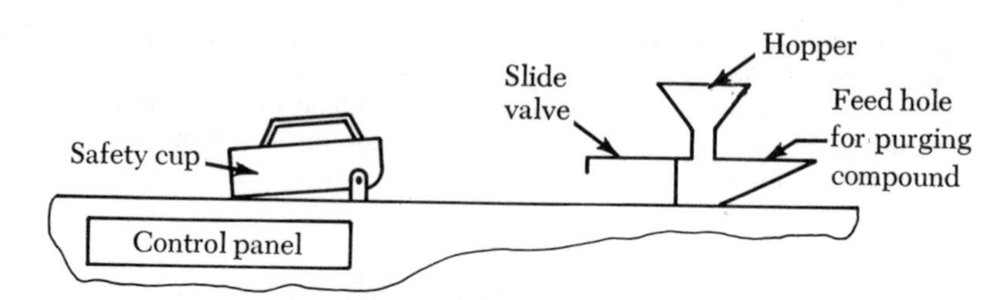

Figure 3. Right-side view of molding machine.

4. Pour 1 cup of purging compound into the hole just below the base of the hopper (Figure 3). The compound and the measuring cup are on the right at the back of the machine.

5. Return to the front of the machine and close the safety gate to allow operation of the machine. NOTE: The mold should stay open during the entire purging procedure.

6. Pick up the right-hand asbestos glove lying on the work table adjacent to the machine. IMPORTANT: You must wear this glove to prevent burning your hand when holding down the safety cup and, later, removing the scrap plastic.

Purging

7. Return to the control panel. Push, but do not release, the fluid meter button until the screw has stopped moving to the right.

8. With the right hand, and with the asbestos glove on, pivot the safety cup over the nozzle hole and hold it down firmly (Figure 3).

9. With the left hand, push, but do not release, the injection and nozzle buttons at the same time. When the screw is no longer moving to the left, release both buttons. CAUTION: Be sure to hold the safety cup down firmly to prevent the hot plastic from squirting on you.

10. After releasing the injection and nozzle buttons, use the right hand to pivot the safety cup to the right. With the gloved hand remove the scrap plastic from the nozzle area and put it in the waste barrel. CAUTION: The plastic is hot, but you will not be burned if you wear an asbestos glove when handling the plastic.

11. Repeat steps 7-10 until all of the purging compound has disappeared from the feed hole and is in the barrel. After step 10 when the purging compound is gone, push the slide valve below the hopper to the left.

12. Continue with steps 7-10 until the grayish color, due to the purging compound, is gone, and the original color of the plastic has reappeared. This return of color signifies that all of the purging compound has been removed and the barrel and screw are ready for normal molding.

Resetting

13. After completing step 12, raise the safety gate and switch the control button to automatic. Be sure that the safety cup is to the right and does not cover the nozzle hole. If it does, it will be crushed when the press closes.

14. After repeating only step 7, return to the front of the machine and continue normal molding operations.

* * * * *

Following each of the steps described above and purging frequently should help in making flaw-free parts. The procedure is quite safe if you are careful, especially during steps 8-10.

—R. R. Nolte

12

Organizations and Layouts

Descriptions of organizations and layouts[1] are types of analysis common in scientific writing. They are typical analyses in that they break down a complex subject into its elements according to a stated principle and arrange the parts systematically. The organization might be a manufacturing company, a bank, a firm of lawyers, a university, a department store. The number of employees might range from a few to many thousands. The layout might be that of a factory, a store, a garage, a farm, a junk yard. The variety of possible subjects for both types of analysis is almost endless.

DESCRIPTION OF AN ORGANIZATION

An organization is a body of people who are mutually dependent on each other but have different functions and duties. As a whole the group has a function or purpose, but each of its divisions or departments, each of its individual members, has a special work to do. The description should explain what the organization does, what the responsibilities of each group and each individual employee are, what their relations to each other are, and how the group as a whole operates.

Plans for Description of an Organization

Two plans for describing an organization are outlined below. Specific subjects may require some modification of these plans, or perhaps quite different plans. Different audiences may also require different plans. Describing an organization's structure for new blue-collar employees may well require a different plan than describing it for potential management-level employees.

Plan One

I. Introduction
   A. Functional purpose of organization
   B. General description
      1. Number of supervisors and workers
      2. Nature of work
      3. Location
   C. Order of reporting
      1. Levels of authority
      2. Responsibilities of supervisors
      3. Duties of workers
      4. System in operation

Plan Two

I. Introduction
   A. (Same as in Plan One)
   B. (Same as in Plan One)
   C. Order of reporting
      1. Levels of authority
      2. Description of organization by divisions
      3. System in operation

Plan One

II. Levels of authority
   A. Highest level of supervision
   B. Next highest level, etc.
III. Responsibilities of supervisors
   A. To other supervisors
   B. To workers
IV. Duties of workers
   A. Classification into groups
   B. Techniques required of each group
V. System in operation
   A. Main steps
   B. End product
VI. Conclusion
   A. Summary of the organization system
   B. Place of the organization in the over-all plant system

Plan Two

II. Description of organization by divisions
   A. Group One
      1. Duties of workers
      2. Techniques required
      3. Responsibilities of supervisor
   B. Group Two, etc.
III. System in operation
   A. Relationships of the divisions
   B. Relationships of the supervisors and workers
IV. Conclusion
   (Same as in Plan One)

INTRODUCTION

At the beginning make clear the specific function of the organization or department you are describing and also the number of supervisors and workers employed. Summarize the work of the group, and describe briefly the place where it works. At the end of the introduction explain—if it seems desirable—the principle of analysis and the order of topics in what follows.

MAIN DIVISIONS

PLAN *I*.—Explain the sequence of authority in the organization. Each person in a group is answerable to someone on a higher level. You can indicate this sequence in a list, and you can construct an organizational chart that will represent the relationships graphically.

An analysis of the responsibilities of supervisors will indicate their relationships to others. How do they promote the work of the organization? Do they train workers? Do they lend assistance?

In describing duties and responsibilities of workers, divide them into groups and explain the techniques required of each group. Mention also any tools used and materials worked upon. List chronologically the steps in the work of each group.

PLAN II.—Emphasize the division of the organization into its separate groups. After describing the duties and responsibilities of supervisors and workers in each group, explain how they function as a unit.

CONCLUSION

In the conclusion make whatever general statements the preceding discussion seems to call for. These may concern the distinguishing features of the organization, its effectiveness, or—if it is part of a larger organization—its importance and functions in the whole system.

Use of Illustrations

Two illustrations may supplement the description of the organization. An organizational chart will show the levels of authority from the top supervisor to the workers. A flow diagram will help the reader to understand the sequence of steps in its operation. Decide whether both are needed. Number and title each chart and diagram, and refer to it by number and explain it in your analysis. Remember that the complexity and amount of detail in illustrations, just as in your writing style, should be shaped by your audience's needs. Flow diagrams and organizational charts, like any other visual, need to clarify things, not confuse them.

An organization chart is reproduced below (Figure 12.1), followed by a flow diagram (Figure 12.2).

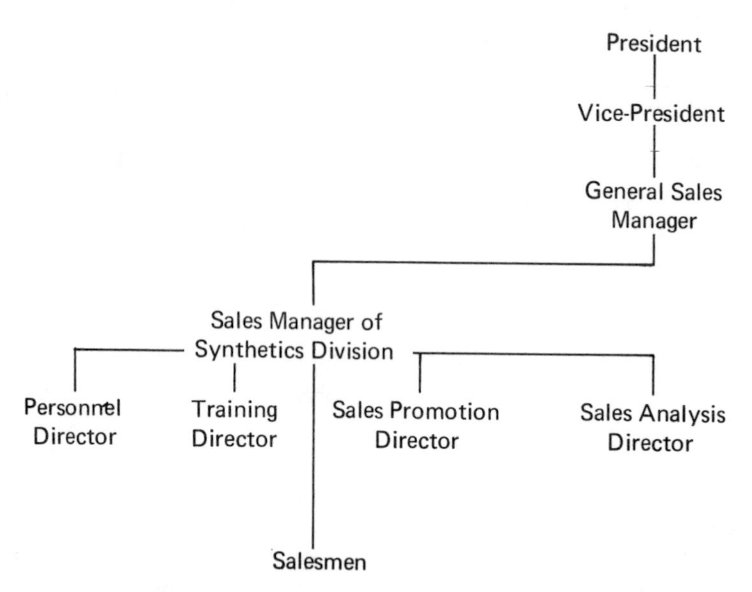

Figure 12.1. Organization of the synthetics division, Firestone, Inc.

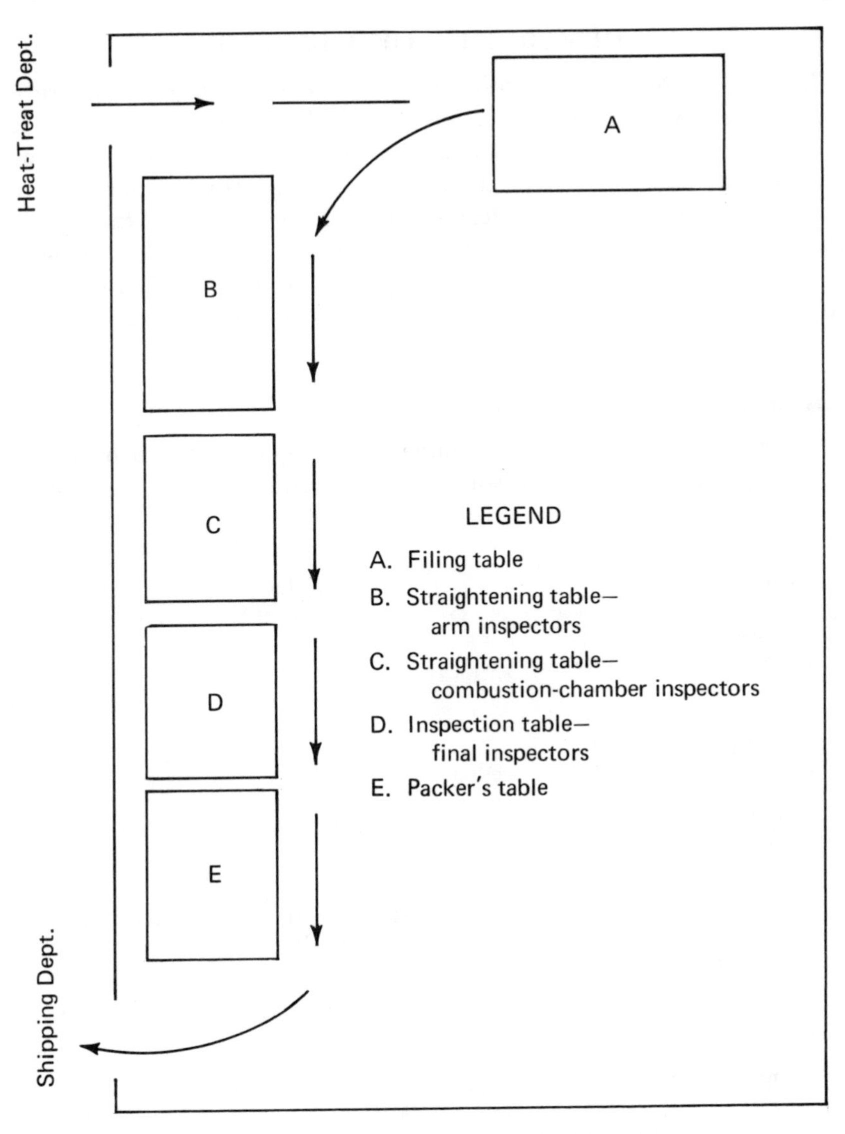

Figure 12.2. Flow diagram—Inspection Department.

DESCRIPTION OF A LAYOUT

"Layout" is a term used to designate the arrangement of machines, furnishings, etc. in a factory or building. In any large-scale manufacturing organization, the industrial engineer is frequently called upon to make recommendations for improving a layout, to give instructions to people responsible for improving it, or to direct their attention to its critical features. The description of the layout will usually be subordinated to one of these purposes. While this purpose will undoubtedly shape your description, certain stages are basic to nearly any scientific description. Those stages are described in the following section.

Plans for Description of a Layout

The outlines which follow may guide you in organizing your own description, but remember that specific subjects may require variants of these plans.

Plan One

I. Introduction
   A. Statement of purpose (if needed)
   B. Statement of limitations (if needed)
   C. Location of layout
   D. Purpose of layout
   E. Plan to be followed (if needed)

II. Item 1
   A. Physical description
   B. Functional description

III. Item 2, etc.
   A. Physical description
   B. Functional description

IV. Advantages and disadvantages of layout

V. Ending (if needed)

Plan Two

I. Introduction
   (Same as in Plan One)

II. Physical description of items
   A. Item 1
   B. Item 2, etc.

III. Functional description of items
   A. Item 1
   B. Item 2, etc.

IV. Advantages and disadvantages of layout

V. Ending (if needed)

Introduction

In the introduction make any general statements needed before you take up the first division of your analysis. Your purpose or controlling idea should be either stated or clearly implied. If you are limiting your analysis to certain aspects of the layout, state clearly what the limitations are. Describe the location and state the purpose of the layout if you have not already done so. If necessary for clearness, inform your reader of the principle of division and the plan of the analysis to follow.

You may need other general statements than those listed in the plan. Try to judge the adequacy of your introduction by asking, "Will my reader now understand fully what I am doing and how I plan to do it?"

Physical and Functional Aspects

In a well designed layout every item has both physical and functional aspects. Plan One takes up the items one at a time in some logical order, and considers both the physical and the functional aspects of each. Plan Two takes up first the physical aspects of all items and then their functional aspects. For your particular problem you can choose between these two plans or perhaps devise some combination of them.

If, for example, you are describing the general layout of the small manufacturing plant represented in Figure 12.3, you would (following Plan One) describe the general arrangement of the plant as indicated by the letters and labels of the figure. Three divisions seem most suitable:

I. Office and lounge facilities near the entry
II. Manufacturing facilities
  A. Receiving area
  B. Raw materials storage
  C. Machining department
    1. Tool room
    2. Storage for work in progress
    3. Assembling department
III. Storage and shipping facilities

Since the description is not intended to be complete and detailed, the furnishings and arrangement of each area are not specified. Plan Two would probably not be suitable for such generalization.

Advantages and Disadvantages

Your next step is to point out the advantages and disadvantages of the layout described. To do this well requires good judgment and an intimate knowledge of the operation of a department. In the foregoing example the arrangement permits the flow of new material from the receiving area through storage to the machining department and thence to assembly and packaging, and finally to the storage of the finished goods and packaging. The sequence of letters in Figure 12.3 and the routing line indicate, first, areas outside of production (A, B, and C) and, second, the flow of material from receiving to shipping (D, E, F, H, I, J, K), with the tool room (G) an adjunct to the machining department. Since the block diagram indicates only relative locations of areas, it cannot be criticized for providing too little space for some departments.

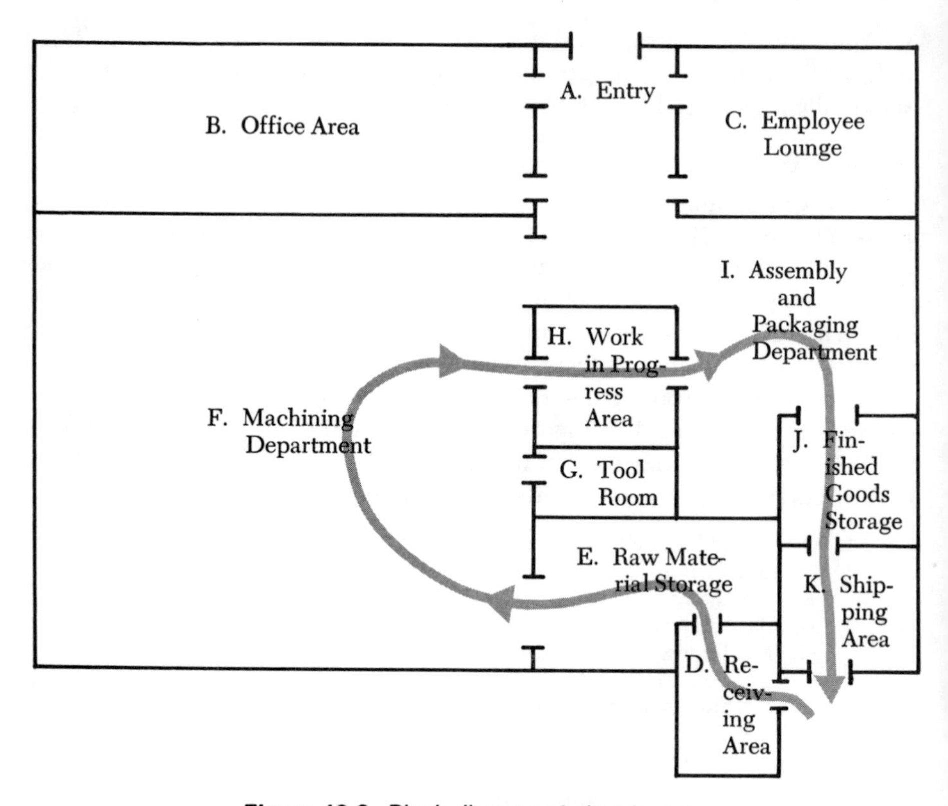

Figure 12.3. Block diagram of plant layout.

ENDING

If the listing of advantages and disadvantages does not provide a good ending, add whatever general statements seem appropriate. Leave your reader with some significant statement about your subject as a whole rather than with some unimportant detail.

Use of Illustrations

To supplement your description include a schematic drawing of the layout where it will be most convenient. Number the drawing, give it a suitable title, and in your description make whatever explanations of it are needed.

EXERCISES

A. *Write a description of an organization. As a subject, choose a small organization such as a factory, a garage, a store, or a department of a large organization. Submit a topical outline with your description. Use headings to indicate the divisions of your subject. Include whatever drawings are needed to supplement your analysis.*

B. *Write a description of a layout. As a subject, choose a small factory, garage, store, laboratory, etc., or part of a larger plant with which you are familiar. Submit a topical outline with your description. Use headings to indicate the divisions of your subject. Include whatever drawings are needed to supplement your analysis.*

C. *Note the organization, the headings, and the illustrations in the following description of a department:*

A GENERAL ACCOUNTING DEPARTMENT

The general accounting department of a corporation is responsible for the handling of accounting information about financial transactions with people or with other companies. Whenever the company purchases or sells any type of merchandise, pays wages, or makes rental payments, it is the job of this department to keep an accounting record of the transaction.

Organizational Structure

One supervisor, the chief accountant, oversees the department of general accounting. The department is divided into five subdepartments: payroll, accounts payable, accounts receivable, billing, and general ledger. The number of workers in each subdepartment will vary with the size of the firm. Since it is the center of the company's financial activity, the general accounting department is usually located at the firm's home office.

Lines of Responsibility

As shown in Figure 1, the general accounting department is under the authority of the controller division. The controller is directly responsible to the president and ultimately to the board of directors of the company. Under the direction of the controller, the chief accountant has authority over the five subdepartments, each described separately below.

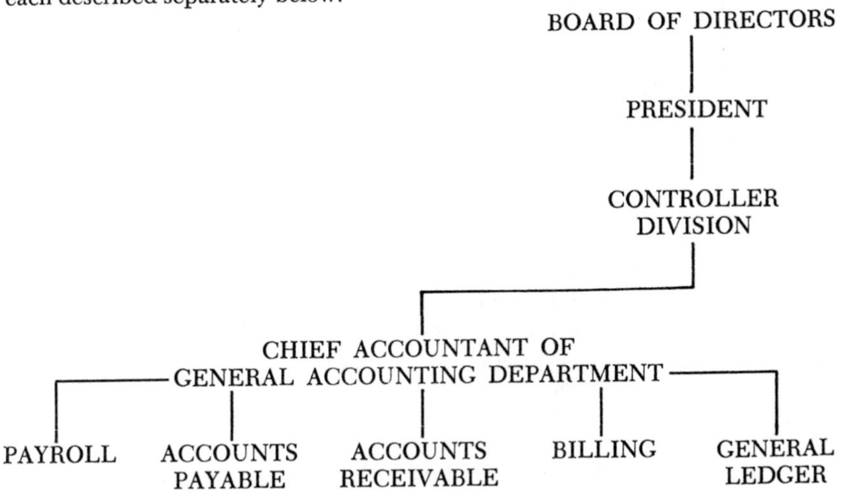

Figure 1. Organization of General Accounting Department.

Duties of Each Subdepartment

The general accounting department is divided into five subdepartments, each responsible for specific duties.

Payroll

The payroll department records and keeps up to date the information necessary to prepare the employees' paychecks. It keeps track of the employees' names, the number of hours worked, and the appropriate wage-rate schedules. After the information is recorded on permanent records, it is forwarded to another division where the paychecks are printed.

Accounts Payable

An accounts payable department is necessary because of the company's willingness to make its purchases on credit. The subdepartment's job is to keep the records of money the company owes to others. Each time a purchase is made on credit, information about the transaction is forwarded to this subdepartment for recording. When the bill is paid, the amount due is subtracted from the accounts payable records. To maintain a high credit rating for the company, the department informs management of accounts payable that would soon be overdue if not paid.

Accounts Receivable

Since the company extends credit to its customers, the job of the accounts receivable department is to keep records of all sales made on credit. When a customer pays for a purchase, the amount owed is deducted from the accounts receivable records. The chief accountant is notified of all accounts that are past the payment-due date.

Billing

After receiving information from the accounts receivable subdepartment, the billing subdepartment prepares and mails monthly notices indicating how much the customer owes.

General Ledger

The general ledger subdepartment is responsible for showing in the accounting records all transactions in which the company took part. The types of information recorded are as follows: the category in which the payment or receipt belongs, the parties involved, the quantity, the dollar amount, and the purpose of the transaction. These records are necessary to upper management, such as the board of directors, which must evaluate the financial position of the company periodically.

Operation of the Department

The operation of the general accounting department is vital to many other departments. Because of the information that it can provide, its main concern is to serve others within the company.

Chief Accountant

The chief accountant is responsible for coordinating the work of the whole department. Since its function is the recording of financial information, the chief accountant makes sure that the records are readily available to people outside the department.

Employees in the Subdepartments

The average worker in a subdepartment is only semiskilled. Once information is provided by other departments or subdepartments, the worker follows a set procedure in keeping the records. When evaluating the output of subdepartments, the chief accountant has authority to change the procedure if it seems advisable.

* * * * *

The general accounting department is an integral link in the flow of information within the company. Though situated down the chain of command, it is looked to constantly for information about the company's financial health. Without its records, the company would have no basis for making important business decisions concerning the future.*

—Dennis Taylor

ORGANIZATION OF THE COURSE "ELEMENTARY SCHOOL INDUSTRIAL ARTS"

The primary purpose of having an organizational structure for the course "Elementary School Industrial Arts" is to establish uniformity of instruction. Though such a structure cannot insure complete uniformity of teaching in all sections of the course, it will provide better directions without altering the traditional plan of the course.

The reorganization is indicated in the block diagram of Figure 1.

Large-group instruction (A) consists of a lecture one period each week for all students enrolled in the course. The viewing of films, giving of demonstrations, and taking of tests are also parts of the instruction for the whole class.

Small-group instruction (B) consists of individual sections, each of approximately 24 students. These meetings are devoted to questions arising in the large-group classes and to many other activities, including (1) lectures and demonstrations by the section instructors, (2) observation of the laboratory school, (3) manipulative activities, (4) reports, and (5) section-level tests.

The plan, as indicated by the dotted-line block C, allows for additional sections if they are needed.

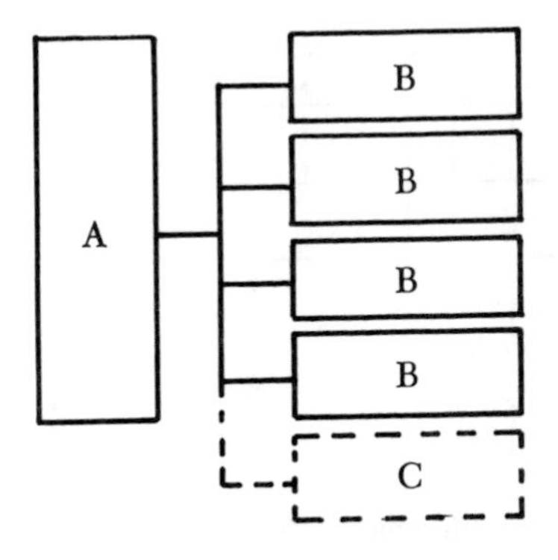

Legend

A. Large-group instruction for all sections
B. Small-group instruction within individual sections
C. Additional sections added as demand indicates

Figure 1. Instructional format for "Elementary School Industrial Arts."

*The facts in this paper were derived partly from Howard F. Stettler's *Systems-Based Independent Audits,* Prentice-Hall Inc., 1967.

RESPONSIBILITIES OF PERSONNEL

Each level of the organization requires an individual or group of individuals at its head to insure progress in instruction. Structurally similar to Figure 1, Figure 2 assigns the specific responsibilities to those in charge of the various sections.

The Coordinator. The coordinator (X) is responsible for making the final decisions about the form of the large-group instruction. He or she is also responsible for drawing on the individual talents of the section instructors (Y) to present materials during the large-group instructional periods. It is his or her job also to monitor the entire procedure to assure as much consistency of work among sections as possile. Authority to perform these duties comes from the chairman of the department.

Section Instructors. The section instructors have a number of duties and responsibilities:

1. Attending all of the large-group meetings.
2. Helping to formulate policies of the course.
3. Delivering lectures in their areas of specialization to the large-group meetings. (The broken lines of Figure 2 symbolize this responsibility.)
4. Instructing the sections assigned to them.
5. Reporting regularly to the coordinator.

DUTIES OF PERSONNEL

Ideally, the entire organization works as a team to train the students. Since the coordinator (X) is often a section instructor as well, instructional duties are common to all. The methods of teaching necessarily vary from section to section, depending on the thinking of the one in charge. The primary duty of the individual instructor, however, is to attain the objectives established and agreed upon by all of the staff.

In addition to performing duties, the coordinator functions alone as the organizer of the course as a whole. The judgment and the recommendations of the instructors make the total program meaningful.

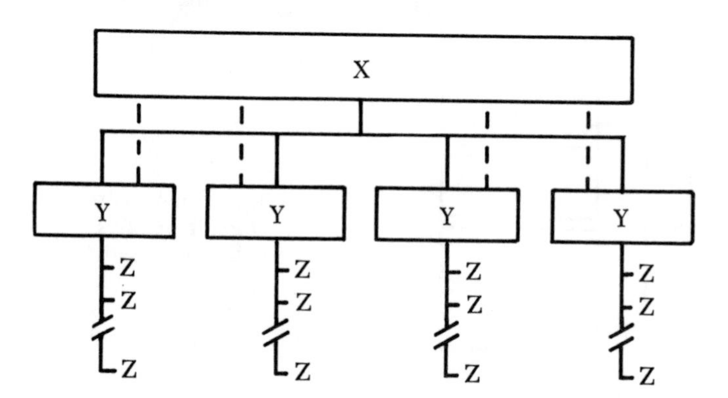

Figure 2. Personnel organization for "Elementary School Industrial Arts."

THE SYSTEM IN OPERATION

Before the beginning of any quarter, the coordinator calls a meeting of all persons assigned to teach one or more sections of "Elementary School Industrial Arts." During the meeting the staff agrees upon any changes to be made to improve past performance. The coordinator then announces the calendar of presentations for the large-group meetings and arranges to make the necessary reservations for outside speakers, visiting professors, audiovisual equipment, testing materials, etc.

Meanwhile the instructors formulate their plans for conducting the work of their own sections. If tasks have been assigned to them in the large-group meetings, they also plan for those.

Each student in the course is assigned to a particular section meeting three hours per week for small-group instruction. One common hour per week (in addition to the three hours for section meetings) is scheduled for all. Typically this meeting is in a large auditorium, with the coordinator directing the program.

EVALUATION

After completing the course the students will be rated for (1) their performance in the testing program of the large group and (2) their performance in the sections. The testing program for the large-group meetings is the responsibility of the coordinator, in consultation with the section instructors. The ratings on the section levels are the responsibility of the section instructors. Through minor changes will occur from quarter to quarter, the typical ratio of large-group and small-group grades in arriving at a final course grade is one to two (1:2).

JUSTIFICATION

The technique of using both large-group and small-group instruction in a single course allows for the freedom of the instructors to handle individual differences, and yet maintain uniformity of progress among sections. Though differences of emphasis in subject matter and teaching techniques are maintained, a common base remains by which students know that the course is a whole and not a number of unrelated parts. Also the instructors are better able to strive for the same educational goals when there is a common element of large-group instructors.

The receptiveness and effectiveness of the coordinator are essential to the overall functioning of this structure. The coordinator must encourage and advise the section instructors, who presumably are less highly motivated. Highly motivated instructors are enthusiastic. Their enthusiasm is transferred to the students, who are motivated to learn.

—William J. Lacroix

Notes

1. The material of this chapter is largely quoted or adapted from leaflets prepared by the Department of English, General Motors Institute, with the permission of C. A. Brown.

13

Abstracts

As the term is used in scientific writing, an abstract is a condensed statement of the most important ideas in a long paper or report. The abstract may vary in length from a single sentence to several paragraphs, depending on its purpose and the nature and length of the original paper. The word is derived from the Latin *abstractus*, from the verb *abstrahere*, to draw from or separate.

Other terms with about the same meaning are *summary, epitome, synopsis,* and *abridgment*. A summary, however, is a part of the original paper, usually a conclusion recapitulating the substance of a preceding discourse. An *epitome,* from the Greek meaning to "cut into" or "cut short," differs very little in meaning from *abstract,* though it is less frequently used in scientific writing. A *synopsis* is an ordered arrangement of main points, perhaps in outline form, that can be apprehended at a glance. An *abridgment* is a shortened form of the original. *Reader's Digest* and similar publications usually publish abridgments rather than abstracts of original articles and books. In an abstract or epitome, however, though there is no objection to the use of the original wording, it will usually be more economical to express the main points in a slightly different wording.

IMPORTANCE OF ABSTRACTS

Before 1900 the amount of scientific material being published in most fields was comparatively small. Workers could keep abreast of new knowledge by reading all papers as they were printed. Since 1900, the number of professional journals has increased to the point where workers can keep abreast of developments only by reading abstracts. To read all of the articles published in one's field would take all of any professional's time.

In spite of concerted and widespread efforts, the problem of information retrieval is far from being solved. Many plans for using computers and combining and systematizing the many abstracting services have been proposed. Research toward documentation, machine searching, and machine translating is continuing, and eventually we may have a center where, as Chancellor Stafford L. Warren of the UCLA medical school has dreamed, "every piece of published scientific literature is reduced to a code and scored on computer tape." All the researcher will have to do is to program the needs, and the computer will instantly respond with an up-to-date record. But until that time comes, the researcher will have to dig out the information needed from indexes and abstracts. While the growing number of computerized indexes has made it somewhat easier, the task of making a complete search of the literature on any subject is still formidable. For a partial list of the principal abstracting journals and computerized indexes, see the section on Sources of Information in chapter 18 and the appropriate section of the bibliography at the end of this book.

Because abstracts provide the quickest access to the mountains of printed material in any field, the student or professional needs to know what an abstract is and how to go about writing one. A well-written abstract not only makes your report more accessible and more attractive, it may also make it easier for readers to understand by providing a quick summary of the contents, which then guides the reader's understanding through the report.

SUBJECT MATTER

You should limit your abstract to a restatement of the important ideas of the original. You are not a critic, remember; it is not your business to give your opinions about the material to be abstracted. Even though you may disagree with the conclusions, you should limit the abstract strictly to summary. The book review, in its criticisms, goes beyond the abstract.

USES OF THE ABSTRACT

Abstracts have many uses. When printed with the original paper or report (usually at the beginning), they provide a "preview" of the content, from which readers can determine quickly whether they should read the whole. For readers with little time they make a reading of the whole unnecessary.

Such author's abstracts are also, as the editors of *Industrial and Engineering Chemistry* suggest, "of service to abstractors," and "facilitate preparation of proper and adequate index references."

When printed in periodicals separately from the original, they enable the busy person to keep informed of recent developments and to sift from many publications those of immediate concern and importance. In most fields of research, abstracts of current literature are now regularly made available in the journals.

LENGTH OF THE ABSTRACT

The length of the abstract will ordinarily be specified by the editor of the journal in which the report is to be printed. In some journals only a one-sentence abstract of a paper of 2,000 or 3,000 words will be required; in others, several hundred words may be used. Note the length of abstracts published in your field and conform to the standard practice. Probably the average length will run from 1 to 3 percent of the original, or from 50 to 150 words for a 5,000-word paper.

If your abstract is separately printed from the original, you can determine the length. Usually it shouldn't exceed about 3 percent.

PROCEDURE IN WRITING AN ABSTRACT

If you are abstracting your own paper or report, some of the steps listed here may be omitted or shortened, for your familiarity with the material may enable you to proceed immediately with the first draft.

Reading the Article

Read the report straight through to get a bird's-eye view of the whole and to make out the author's main purpose.

Noting Main Facts

Determine the main sections of the original, and identify the key sentences. If the article is well written, with proper transitions and "signposts," this step will be easy. If it is poorly organized, with unmarked transitions, abstracting may be difficult.

Making an Outline

Make a working outline, including the main points with their divisions and perhaps subdivisions. Do this to avoid omitting or wrongly emphasizing points in the abstract.

Writing the First Draft

Write the first draft, with all the points in your outline stated in complete sentences. It should be a "scale model" of the original, with essentially the same proportions.

Begin with a statement of the central idea. Then summarize all important points, in the same order and with the same proportions. By omitting important points or overemphasizing minor ones, you could give a wrong impression of the original.

At this stage make no special attempt to economize on the number of words. Do that later.

Expanding or Condensing the First Draft

If the first draft is too short, check minor points not covered before, and expand the first draft to meet the requirements. Probably, however, the first draft will be too long. Reexamine the outline to decide whether you should omit subordinate points. If, for example, the outline has divisions as follows,

1. Main division
    A. Subdivision
        1. Section
        2. Section
    B. Subdivision
        1. Section
        2. Section
        3. Etc.

consider whether you should drop points under A and B.

Attempt final condensation by thorough sentence revision. Often you can combine statements of coordinate value by the device of parallel structure. Often you can condense clauses to phrases and phrases to words.

Don't use "telegraphic style" to economize, however. Include the subjects, verbs, articles, prepositions, etc. Instead of "sighted sub, sank same," write, "I sighted a submarine and sank it."

In brief abstracts, especially those found in annotated bibliographies, incomplete predications are used to save space. The following example from *Scholarly Books in America,* July 1964, illustrates this practice:

> DUNNE, Howard W., ed. *Diseases of Swine.* 2nd ed. 807 pp.
> 7 x 9⅝ in. $14.50. Iowa State, 63-16675
>
> A professional one-volume encyclopedia of swine disease information by 56 contributing authors. Features black and white and colored photomicrographs, newest drugs and dosage ranges, and latest information about current research.

Use suitable transitional words and phrases to make the finished abstract read like a connected and unified whole, not like a series of disjointed and unrelated statements.

Supplying Bibliographical Data

Supply, usually at the beginning, the name of the author of the article, its title, the publisher, the date and place of publication, and any other data necessary for complete identification.

TWO FORMS OF ABSTRACT

Two forms of abstract are in common use:

(1) The *descriptive abstract,* which describes the paper or report, tells what it is about. Often it is a section-by-section epitome, like this abstract of a long essay by Sir J. Arthur Thomson:

> Sir J. Arthur Thomson, "The New World of Science," *The Atlantic Monthly,* June 1930, vol. 145, pp. 838-850. This paper, summarizing revolutionary scientific discoveries marking steps in man's progress, has four main sections:
>
> Section I, "Retrospective," lists significant early discoveries that made the world appear in a new light.
>
> Section II, "The World Today," describes four recent advancements resulting from the development of scientific unifying ideas.
>
> Section III, "The Old Giving Place to the New," lists forces which, because of these advances, no longer intimidate us.
>
> Section IV, "New Prospects," considers some of the fields in which future developments may bring additional transformations.

(2) The *informative* abstract, which summarizes the most important points of the original. An informative abstract of the same essay by Sir J. Arthur Thomson might read like this:

> Sir J. Arthur Thomson, "The New World of Science," *The Atlantic Monthly*, June 1930, vol. 145, pp. 838-850. This paper summarizes the revolutionary scientific discoveries marking steps in man's progress. Significant early discoveries were those of the order in nature, the heliocentric principle, the law of gravitation, and the principle of the conservation of energy. Recent advancement has been marked by the establishment of four significant unifying ideas: (1) that all matter is basically the same; (2) that radiant energy is universal; (3) that all organic life is cellular and protoplasmic; and (4) that mind pervades all animal life. Forces that formerly intimidated man no longer operate: the notion of forces leagued against him, the sense of confusion, the bogey of the capricious, the fear of illegitimate materialisms, the absence of any hope of progress, the fear of ineradicable evils, the false hope that science will solve all problems. New prospects are the elimination of disease, improvement of the race, progress through education and social planning, and emancipation of mind from matter.

Combinations of the descriptive and the informative forms are common. In general, the shorter the abstract the more likely it is to be descriptive. The abstract of the chapter on Mechanical Aspects of Style which follows is both descriptive and informative.

Since the descriptive abstract tends to be merely an outline of the original or a list of topics covered, the informative abstract is usually to be preferred.

EXAMPLE OF PROCEDURE IN PREPARING AN ABSTRACT

Following are shown three stages in the writing of a 200-word abstract of chapter 19 in this book, Mechanical Aspects of Style: the outline, the first draft, and the final draft. The length arbitrarily specified is about 2 percent of the length of the chapter.

The *combination form* of outline (see chapter 2 on kinds of outline), in which main items are complete predications and less important ones are in topical form, is used.

Controlling purpose: to describe current practices in some mechanical aspects of style in which usages may vary.

I. *General statements*: A knowledge of usages governing mechanical aspects of style is essential for writers who would turn out good typed manuscript.

   A. Mainly aspects of style in which usage is divided are analyzed here.

   B. Established usage in their own fields should be observed by writers and adhered to constantly.

   C. Though there are many standard reference works on usage, the general statements following are based upon studies of actual usage in scientific writing.

II. *Specific aspects of usage*: Practices in six different categories are summarized in the following sections:

A. The list of standard abbreviations recommended by the ASA (American Standards Association) is reprinted, and conclusions are reached about:
1. The tendency away from excessive use of abbreviations.
2. The dropping of capital letters and periods.
3. The use of one form for both singular and plural.

B. The reasons for the hyphenation of compound adjectives are given, with special attention to:
1. Exceptions to general practice.
2. The suspension hyphen.

C. Current practices in the writing of numbers are analyzed, with special reference to:
1. Exact numbers, whether quantitatives or aggregates.
2. Several special cases:
   a. Approximations.
   b. Large numbers.
   c. Numbers beginning a sentence.
   d. Two numbers coming together.
   e. Fractions and mixed numbers.
   f. Decimals.
   g. Units of measurement.
   h. Ordinals.
   i. Small aggregates in figure form.

D. Established practices in capitalization are described for:
1. Titles, heads, subheads, and legends.
2. Outlines and tabulations.
3. Tables of contents.
4. Common nouns used with date, number, or letter.
5. Derivatives of proper nouns.
6. Expressions after a colon.

E. The various methods of indicating titles of books, magazines, etc. are described.

F. The prevailing usages in punctuation, with variants, are described for:
1. The period.
2. The question mark.
3. The exclamation point.
4. The comma (a) to separate and (b) to set off.
5. The semicolon.
6. The colon.
7. The dash.
8. The apostrophe.
9. Ellipsis periods.
10. Quotation marks.

III. Exercises requiring the application of rules of styling are included in most of the parts of section II.

The *first draft* of the abstract, which merely converts the outline into paragraph form, is 286 words long.

This chapter of the textbook *Writing Scientific Papers and Reports* describes current practices in some of the mechanical aspects of style where usages may vary. Writers who would turn out good typed manuscripts must have a knowledge of the acceptable variants and adhere consistently to established usages in their own fields. The general statements made in the chapter are derived not from reference works, but from studies of actual usage.

Acceptable practices in six different categories are summarized in the chapter. (1) The ways of writing abbreviations are analyzed, and the writer is advised to use the forms specified by the American Standards Association, whose list of abbreviations is reprinted for the writer's guidance. (2) The reasons for hyphenating compound adjectives before a noun are given, with special attention to exceptions and to the suspension hyphen. (3) Current practices in the writing of numbers are analyzed, with attention to exact numbers, whether quantitatives or aggregates, and to several special cases involving approximations, large numbers, numbers beginning a sentence, two numbers coming together, fractions and mixed numbers, decimals, units of measurement, ordinals, and small aggregates in figure form. (4) Established practices in capitalization are described for titles, heads, subheads, and legends; outlines and tabulations; tables of contents; common nouns used with date, number, or letter; derivatives of proper nouns; and expressions after a colon. (5) The various methods of indicating titles of books, magazines, etc. are described (6) The prevailing usages in punctuation, with variants, are summarized for the period, the question mark, the exclamation point, the comma, the semi-colon, the colon, the dash, the apostrophe, ellipsis periods, and quotation marks. Many examples of and exercises on most of the points of styling described are included.

The *final draft,* 150 words long, necessarily drops many of the details indicated by Arabic numerals and lowercase letters in the outline. Complete bibliographical data, not counted as a part of the 150 words, are always put at the beginning.

W. Paul Jones, "Mechanical Aspects of Style," from *Writing Scientific Papers and Reports,* Wm. C. Brown Company Publishers, Dubuque, Iowa, Eighth Edition revised by Michael L. Keene, 1981, pages 281-312. This chapter is limited to analyses of mechanical aspects of style where usages vary. Writers should know what the acceptable variants are and adhere to established practices in their own fields. The author's general statements are derived from studies of actual usage. Practices in six different categories are summarized: (1) Preferred practices in writing abbreviations are described, and the American Standards Association list of abbreviations is reprinted for the guidance of writers. (2) The reasons for hyphenating compound adjectives are given. (3) Current practices in writing numbers are analyzed, with attention to exact numbers, both quantitatives and aggregates, and to several special cases. (4) Variant practices in capitalizing titles, headings, outlines, tables of contents, etc. are described. (5) The various methods of indicating titles are listed. (6) The prevailing usages in punctuation for all the marks are described. All general rules are illustrated and most of the sections include exercises.

EXERCISES

A. *Write an abstract of an article assigned by your instructor. With your completed abstract submit your outline of the article and your first draft of the abstract.*

B. *Write an abstract of an article from a recent issue of a journal in your own field. Submit the article or a Xerox copy of it with your completed abstract, your outline of the article, and your first draft.*

C. *Note the typical characteristics of abstracts in the following examples from scientific journals:*

1. From *Chemical Abstracts*:

Modification of the fluorescence attachment for the Beckman model DU spectrophotometer. Morris Rockenmacher and Andrew F. Farr (Univ. of California, Los Angeles). *Clin. Chem.* 9(5), 554-6(1963). The fluorescence attachment for the Beckman DU instrument was modified so that all the filters are readily interchangeable, and the unit does not have to be removed when the instrument is used as a spectrophotometer. The details of the procedure are illustrated.

W. A. Creasey

New double-beam fluorometer. Milton Laikin (Beckman Instrs., Inc., Fullerton, Calif.). *Rev. Sci. Instr.* 34(7), 773-7(1963). A double beam system alternately irradiates sample and reference. The sample and reference signals are sepd. by a ring demodulator. The reference signal is fed back to the photomultiplier dynode supply and the sample signal to a meter which indicates the ratio of sample to reference. Ultimate sensitivity for quinine is 1 part/billion full scale.

G. J. Alkire

2. From *The Geophysical Journal* of the Royal Astronomical Society:

A GEOPHYSICAL STUDY OF THE RED SEA

C. L. Drake and R. W. Girdler

(Received 1964, February 4)

Summary

A description is given of the bottom topography, seismic refraction profiles, magnetic and gravity observations in the Red Sea. The deep trough along the centre is associated with positive Bouguer gravity anomalies, large magnetic anomalies and seismic velocities of 7-1 km/s. It appears that this represents a fissure in the continental crust, partly filled with basic, igneous material. A structural map, based on all the geophysical evidence, has been prepared and it is suggested that the complex Red Sea rift was formed as a result of crustal tension. Finally, a discussion is given of the Red Sea as part of the world rift system.

Introduction

In 1958 a joint cruise was made in the Red Sea by R/V *Vema* of the Lamont Geological Observatory and R/V *Atlantis* of the Woods Hole Oceanographic Institution. During this cruise 15 seismic refraction profiles were made in the Red Sea and measurements of the total intensity of the Earth's magnetic field, coupled . . . .

3. From *The Journal of the Astronautical Sciences*:

LAUNCH WINDOWS FOR ORBITAL MISSIONS

A. H. Milstead

Abstract

A great many orbital missions involve launching a vehicle into a particular earth-referenced plane. The launch window is defined as the time span around the nominal launch time during which the vehicle may be launched and the target plane achieved within a specified additional ideal velocity budget. This paper presents analytical formulations for the launch window as functions of the additional ΔV budget and other parameters for fixed and for variable launch azimuths. The effect of launch azimuth constraints (e.g., for range safety) on the launch window is investigated and several related problems are discussed.

Introduction

Many of the orbital missions currently under study by the civilian and military space agencies involve the launching of a vehicle from an earth-fixed site and the establishment of that vehicle in a particular plane (hereafter called the target plane) passing through the Earth's center. Such a mission, for example, is the . . . .

4. From the *Journal of Petrology*:

PETROFABRIC ANALYSES OF RHOM AND SKAERGAARD LAYERED ROCKS

by R. N. Brothers

Department of Geology, Auckland University, New Zealand

Abstract

Petrofabric analyses of layered rocks from Rhum have revealed a preferred orientation for felspar in the allivalites and for olivine in the peridotites; a regional petrofabric map of felspar orientation contains a radial pattern which suggests the presence of convection currents during crystal settling. An oriented specimen of Skaergaard ferrogabbro from the margin of a trough band has allowed comparison to be made between a known magma current direction and the preferred orientation of felspar, olivine, clinopyroxene, and apatile crystals in the rock.

Introduction

Many authors have discussed the mechanism for mineral layering in large basic plutons, and in some cases the combined effects of magma currents and gravity settling have been invoked to explain separation of the mineral fractions now found in the contrasted parts of any one rhythmic unit. Only a few petrofabric . . . .

Reports and Proposals

14

Requisites in Report Writing

Kinds of Reports
- Proposals
- Periodic Reports
- Progress Reports
- Examination Reports
- Recommendation Reports
- Research Reports

Distinctive Characteristics of Reports
1. It is written for one person or a limited group.
2. It is in report form.
3. Its outward appearance is distinctive.
4. It has many headings and subheadings.
5. Its language is formal.
6. It is objective.
7. It is as nearly complete as possible.
8. Its language is clear and concise.
9. It is mechanically correct.

Qualifications of a Good Report Writer
- Firsthand Knowledge of a Subject
- Accuracy
- Objectivity
- Ability to Analyze and Generalize
- Ability to Systematize Facts
- Understanding of the Readers' Needs

Exercise

A report is usually the solution of a problem or the answer to a question, supported by facts obtained or verified by the author. It is a special form of scientific prose developed for the needs of science, engineering, and business, usually written in response to an order, request, or authorization, though sometimes submitted entirely on the initiative of the author.

KINDS OF REPORTS

The information in the report may be of many different kinds. It may concern work in progress or completed; it may be the result of examination or analysis of any of a variety of subjects; it may present the results of investigation or research.

A scientific classification of reports according to purpose or subject is difficult because the variations are so many and the distinctions so indefinite. A classification according to field, scope, or time covered is only approximate. Only one, according to length (*short* and *long*: see the next two chapters), is logical. The types named in the headings below may give some notion, however, of the principal differences.

Proposals

Proposals are written offers submitted by someone to do something for somebody else. Since they concern the future rather than the past, they are really not reports according to the definition given above. But since they are likely to be identical in format with the report, and usually mark the beginning of projects that end with the submitting of final reports, they are listed here. (See chaper 17, Proposals).

Periodic Reports

Periodic reports provide information at regular intervals concerning the status of an organization or an activity. Examples are the annual reports of corporations to stockholders and weekly or monthly reports of technicians or heads of departments.

Progress Reports

Progress reports provide information concerning work done, problems encountered, and difficulties overcome while a project is under way. Examples are reports submitted during the construction of a dam or a building, and during a research project.

Examination Reports

Examination reports provide information derived from firsthand knowledge of a subject, usually with conclusions drawn from study of the information. Examples are reports on the condition of a building, a forest area, or the finances of a company, and on the natural resources of a state.

Recommendation Reports

Recommendation reports serve as the basis for future action. Usually they include the facts which support the recommendations, since unsupported recommendations, even from persons of great prestige, are unlikely to be adopted. Examples are reports concerning the selection of sites for factories or stores, the ways of increasing the efficiency of manufacturing methods, and the choice of common stocks for investments of capital. "Feasibility" and "Staff-Study" reports are likely to include recommendations.

Research Reports

Research reports reveal the discovery of hitherto unknown facts derived from experiment, observation, surveys, and accumulated data. University laboratories, engineering and agricultural experiment stations, government bureaus, and private research organizations issue many such reports.

It is obvious that these classifications overlap. The progress report may include recommendations; the recommendation report may result from the examination of facts; research reports may be submitted periodically. The naming of the categories[1] is unimportant. Writers will know what kind of report to prepare when they reach the stage of organizing material.

DISTINCTIVE CHARACTERISTICS OF REPORTS

Reports may differ from other forms of scientific prose in several respects. The first five in the following list are characteristic of most good reports; the last four, of all good scientific writing. (See chapter 1, Some Generalizations.)

1. It is written for one person or a limited group.

Usually you indicate at the beginning to what person or group you are addressing the report. Most frequently you are responsible to the person or group addressed, either as a student, an employee, or a retained adviser, and submit a report in response to an assignment, a request, or an order.

If you have no direct authorization, as happens frequently in experimental or research work, you still address your report more or less directly to specialists. Only occasionally will you address a report, perhaps on a subject of wide public interest like water pollution or soil conservation, to the general reader. When the subject has wide general interest, it is more likely to be published in book or pamphlet form than in report form.

2. It is in report form.

Usually you submit a report in what is known as "report form"; that is, it has a title page, a letter of transmittal, a table of contents, an introduction, conclusions and recommendations, and frequently an appendix. If it is a short report in letter form, its divisions are likely to be indicated by headings.

3. Its outward appearance is distinctive.

Usually your report will be typed, it will have an attractive cover of heavy but flexible material, with the title and your name, perhaps other identifying data, on the outside, and is made up with careful attention to the first impression it will make on your readers. If that first impression is favorable, they will be more inclined to attend to the conclusions and recommendations.

4. It has many headings and subheadings.

Readers rarely read a report through from beginning to end, as they would read a short story. They look for what they want to know, the purpose of the report, an abstract of its substance, especially a list of its conclusions and recommendations, perhaps some section in which they have a special interest—and the headings help them to find it. Even in a short letter or memorandum report, the headings are essential.

5. Its language is formal.

The general style of a good report is formal. Slang and highly colloquial language, even contractions, are acceptable in oral reports, but rarely in written ones. Personal pronouns—"I," "we," "you,"—are appropriate only when the expression of personal opinion or the narration of personal experience is justifiable.

Exceptions may be in letter reports or letters of transmittal and proposals, in conclusions based partly on the author's judgment, and in sections using personal experience and observation as bases of generalizations. Instead of writing "It was observed that the surface of the pavement was corrugated," write, "I observed that . . ." or—better still—"The surface of the pavement was corrugated," leaving the implication clear that the author had so observed. Instead of writing, "It is the opinion of the writer that the asphalt surface had been poured when weather conditions were poor," it is better to write "I believe that . . ." or—again, better still—"The asphalt surface was poured when weather conditions were poor." Any positive statement in the report is a belief of the writer; there is no need to say, "I believe this statement."

6. It is objective.

The purpose of a good report is primarily to present facts. Any conclusions stated should be inductions derived from the specific evidence of the report. Recommendations should exclude mere personal opinion or evidences of personal interest or prejudice. If the data of the report are inconclusive or conflicting, readers should be left in no doubt that any conclusions based on that data are tentative.

7. It is as nearly complete as possible.

The good report leaves no unanswered questions in readers' minds. It includes all the pertinent material. It makes all the pertinent generalizations that can be derived from the facts. If, from unavoidable causes, your knowledge of the subject is incomplete—because of lack of time, lack of money, lack of needed equipment—you indicate what is lacking and why it is lacking.

8. Its language is clear and concise.

The language of your report should be as clear and concise as you can make it. Clearness and conciseness do not necessarily go together. The sentence "Spur gears are the only types of gears to be used" is clear; "Only spur gears are used" is both clear and concise. The original has 11 words, the revision has 5 words. If the sentences of a whole report are so wordy, half its length is wasted, and at the same time the readers' patience has been exhausted.

For further analysis of such weaknesses consult the sections on Wordiness and Clearness in chapter 20.

9. It is mechanically correct.

You cannot afford to submit a report with mechanical faults. A mistake in spelling or grammar will make readers wonder whether a writer so careless would not also be guilty of mistakes in recording data or making calculations. Even an inkblot on a page or a misspelling distracts attention: the mind concentrates on *seperate* and overlooks the fact being reported. Any mechanical fault, in spelling, grammar, sentence structure, punctuation, or misuse of a word, is distracting and effects your readers' confidence in you.

Taking chances on being correct does not pay. A student recently misspelled *convenience* in his report, *convience;* the instructor marked it and the student corrected it, *convienience.* He was too lazy to use a dictionary—like the freshman who spelled *vessel* two or three ways, crossed them out, and substituted *boat,* spelled *bote.* Use your dictionary. A guide to good usage, like H. J. Tichy's *Effective Writing for Engineers—Managers—Scientists* (John Wiley and Sons, Inc., 1966), can be a good investment.

* * * * *

Other characteristics of a good report could be listed, such as the frequent use of visual aids of all kinds, and its "double" form—a condensed version at the beginning for the reader in a hurry, and an extended version in the body of the report for the reader who wants all the facts.

QUALIFICATIONS OF A GOOD REPORT WRITER

Although the characteristics of a good report are listed in the preceding section, it is unfortunately true that you can know what a good report is, but still write a poor one—in the same way you can know a great deal about music, yet still be unable to play even a simple melody on the harmonica. You need certain skills and abilities to enable you to put your knowledge into practice.

Firsthand Knowledge of a Subject

No one could write a good report on the construction of the Alaskan pipeline whose information did not come mostly from direct experience and observation. Of course only a person having that firsthand knowledge would be asked to write a report about it.[2] Only students desperately in need of a subject for a report or term-paper assignment would think of writing about something that they know only from their reading.

Frequently, however, firsthand knowledge will need to be supplemented by knowledge derived from other people's experience. If the subject is a large one, such as the construction of the Alaskan pipeline, even the chief engineer would need to draw upon the knowledge of subordinates, and consult records, maps, and statistics to complete the report.

Accuracy

You need to check and double-check every fact, every figure, every general statement. If readers discover a single mistake, a single wrong statement, their confidence in you may be shaken. If you are in doubt about any statement and cannot make sure, either omit it or state that it is doubtful.

Objectivity

Objectivity is essential for accuracy. As a report writer, you should be like a thinking machine, without the passions and prejudices that are liable to influence you. The facts should justify your conclusions and recommendations, even though those conclusions and recommendations are contrary to your own desires or may result in action to your disadvantage.

Ability to Analyze and Generalize

A report is usually analytical in its method. You take the subject to pieces, display its different parts, and demonstrate their interconnections and relationships. Then you apply the inductive method to arrive at general conclusions. If you generalize on too few data, or ignore data which would not support the conclusion you desire, or if your data do not support the conclusions you want to reach, your report may do more harm than good.

Ability to Systematize Facts

You must be able to arrange the parts of your subject in logical and systematic fashion. If you are using firsthand information, you are likely to have an embarrassment of riches at your disposal. Before you write your final report, eliminate what is not pertinent and systematize what remains. Classify and group your facts and prepare an analytical outline that displays them to their best advantage. Then present them so that your plan and your logic are evident.

Understanding of the Readers' Needs

Since reports are usually prepared for one person or a limited group, you must be sensitive to their needs. What you put into the report, what you omit, what terms you define, what you take for granted, what you illustrate or explain, how you arrange your material—the answers to many such questions depend on your understanding of the reader or readers. The most general principle in report writing is to seek constantly to save the reader work.

The report should be organized in such a manner as to reflect the nature of the reader's needs rather than the writer's. Too often, the structure of a report reflects only the way in which the subject developed in the writer's understanding, the way in which the subject became clear to you as writer. That may be fine for the first draft, but as you revise it you need to arrange it in the way it can most easily be understood by the reader. The scientific report, perhaps more than any other form of scientific writing, needs to be reader-centered, not writer-centered.

* * * * *

These qualifications are not listed in the order of their importance. Some analysts would probably put "Understanding of the Reader's Needs" first and "Objectivity" last. Which are most important will depend on the subject, the objectives, and the reader addressed.

EXERCISES

Look up the definition of "Report" in various dictionaries and reference books and write an analysis of what you find.

Notes

1. To the list might be added the *laboratory report,* which, unlike most reports, is submitted not to provide information or serve as the basis of future action, but to serve as a test of the student's knowledge and skill. Other terms often used in referring to reports are "preliminary, partial, interim, final, completion, status, experimental, special, trade, formal, service, operation, construction, design, failure, student-laboratory, industrial-research, industrial shop, evaluative, test, . . . examination-trip, inspection, investigation, memorandum, note-book, short-form, . . . information, and work."—Gordon H. Mills and John A. Walters, *The Theory of Technical Writing,* Circular No. 22, The University of Texas, Bureau of Engineering Research.
2. Reports summarizing known information are sometimes submitted, but the person asked to prepare such a digest is likely to have wide firsthand knowledge of the subject.

15

Short Reports

Most reports of not more than three or four double-spaced typed pages are submitted in short-report form. Their subjects are too limited, or their treatments not sufficiently final, to justify either the length or the form of the long report.

The two commonest forms of the short report are the letter and the memorandum ("memo"). The memo is likely to be less formal, and lacks some of the elements usually found in the letter report.[1]

While this book does not cover the broad fields of "business letters" or "business correspondence," the form and layout of short reports are the same as those of business letters. See any of the books listed on page 353.

THE LETTER REPORT

In appearance and layout the letter report is like the business letter; it has the conventional heading, inside address, salutation, body, and complimentary close. It may or may not be typed on a printed letterhead. The subject is frequently indicated by a suitable heading, the language may or may not be formal, and the sections of the body are usually indicated by numbers or by headings.

Appearance and Layout

Write the short report on a printed letterhead or on plain 8½- by 11-inch paper, usually a good grade of white bond. If the report is short enough to be written on a single sheet, the layout of parts is exactly like the layout of parts of the short business letter. Center the material on the page, with side margins equal and margins above and below approximately the same. Leave all margins at least an inch wide in typed reports.

If the report requires more than a page, leave a 1-inch margin at the bottom of the first page and begin the second page about an inch from the top, even if there are only a few lines left over. Do not attempt to center the written material vertically on the second page unless it is a full page.

Approximately the same rules govern the appearance and layout of reports typed or written in longhand, except that margins in those done in longhand are usually narrower, about a half-inch all around, and the layout is like that of a double-spaced typed manuscript.

To be professional, your report really should be typed. Use a fresh ribbon, and be sure the typeface is clean. Avoid "strike-overs" and messy erasures. Use correction paper, correction fluid, or a correction ribbon; even so, pages with more than a few errors should be retyped. Do not make corrections with pen or pencil.

Either single or double spacing is permissible, but single spacing usually makes a better-looking page. When you single-space, separate paragraphs by double spacing, and double-space above and below center headings or marginal headings.

The forms of layout in most common use (the pure block, the modified block, the indented, and the combination) and the different usages in punctuation are illustrated in the models on pages 188-194.

Elements of the Letter Report

The elements of the letter report, in order from beginning to end, are

1. Printed letterhead (not always used)
2. Typed heading
3. Inside address
4. Special address—"attention" line (not always used)
5. Salutation
6. Subject heading (not always used)
7. Body
8. Complimentary close
9. Signature
10. Supplement (not always used)

The most common practices in the writing and placing of these elements are summarized in the following sections and illustrated in the models. For variations from these practices and for more detailed descriptions, consult one of the standard textbooks on letter writing.

The form used in military practice, and frequently elsewhere, differs from the form here described. Several different forms of reports, including a modified military form, are described on p. 186 and shown on p. 188 through 194.

Printed Letterhead

The simplest and most dignified printed letterheads include merely the name, address, and zip code number of the author of the report or of the organization that he or she represents. The items are arranged in pyramid form in the middle of the top of the sheet. Letterheads including names of officers and pictures of products and cluttered with advertising matter, though sometimes justifiable for business correspondence, are not suitable for reports.

Typed Heading

If the printed letterhead includes your address, the typed heading includes only the date. Place it a space or two below the lowest line of the printed letterhead so that it ends at the right-hand margin of the body of the letter.

If there is no printed letterhead, include in the heading the complete address of the writer of the report, as well as the date and the zip code number. The complete heading will require from two to four or five lines, depending on the number of items in the address. Place it at the upper right so that its longest line ends at the right-hand margin of the body of the letter. The block form illustrated, with open punctuation, is perhaps most common.

On the printed letterhead the heading is always placed at the same location, regardless of the length of the report. If there is no printed letterhead, the amount of space left above the heading depends on the length of the letter: the shorter the report, the wider the space. Leave the margins above the heading and below the signature approximately the same. If the report fills more than a page, leave margins on the first page of about an inch at top and bottom.

Typical headings are these:

Glendora, Iowa
March 6, 19--

142 Chestnut Avenue
Moravia, Kansas 66000
March 2, 19--

Apartment 43
1900 Maple Terrace
Kansas City, Kansas 66118
March 6, 19--

INSIDE ADDRESS

The inside address includes the complete name and address of the person or organization to whom you are sending the report. Place it at the upper left and make its form and punctuation consistent with those of the heading.

If your report is written on one sheet with printed letterhead, the position of the inside address depends on the length of the report. Place it at least two spaces below the date or as many more as may be desirable to make a well-balanced page.

If there is no printed letterhead, place the inside address as many spaces below the heading as necessary to secure good layout.

Typical inside addresses are these:

Mr. John C. Brown
1940 Ormond Street
Philadelphia, Pennsylvania 19105

Professor James E. Snow
Department of Mechanical Engineering
Centralia College
Centralia, Iowa 50123

Joseph E. French, Superintendent
The Waxley Engineering Corporation
5400 South Ardmore Street
New York, New York 10036

Note that, in accordance with prevailing good practice, no abbreviations (except "Mr.," "Ms.," and "Mrs.") are used in either heading or inside address.

SPECIAL ADDRESS

When you address a report officially to an organization named in the inside address, but wish to make sure that it will reach a certain individual in the organization, add a special address. Begin it with the word "Attention" followed by the name and perhaps the title of the individual, and insert it usually two spaces below the inside address, indented five spaces from the left-hand margin; for example,

Attention: Dr. Donald R. Spruce, Vice-president

Salutation

The salutation is formal and consistent with the inside address. If you are addressing a company or organization, the salutation is always "Gentlemen" or "Mesdames," even though you add a special address naming a particular person. If you are addressing an individual, the salutation will be "Dear Sir," "Dear Mr. Brown," "Dear Professor Snow," but not "Dear John" (too informal) or "Dear Prof. Snow." (Titles like "Professor" and "Doctor" are not abbreviated when used with surnames alone.)

The salutation is usually followed by a colon, though it may be left unpunctuated.

Subject Heading

The subject heading is always helpful to the reader of a letter report. Center it two spaces under the salutation; for example:

Subject: Installation of New Transformer

The word "Subject" is sometimes omitted. Sometimes this heading follows the inside address rather than the salutation; sometimes it is on the same line as the salutation.

Body of the Report

Begin the body of the letter report with an acknowledgment of the request or order which led to the writing of the report, and with a statement of the subject of the report.

Such beginnings tend to become stereotyped and trite. Try to vary your wording occasionally if you have many such reports to write. Here are several different ways that one letter report might begin:

1. In compliance with your request of January 15, 19–, I am submitting this report on the installation of the new transformer at Substation C.
2. As you requested on January 15, 19–, I am reporting on the installation of the new transformer at Substation C.
3. Enclosed is the report you asked for on January 15, 19–, concerning the installation of the new transformer at Substation C.
4. The report you asked for on January 15, 19–, concerning the installation of the new transformer at Substation C, is below.
5. On January 15, 19–, you asked me to report on the installation of the new transformer at Substation C.

6. This letter is my report on the installation of the new transformer at Substation C, as you requested on January 15, 19–.
7. The new transformer at Substation C has been installed, and I am reporting on the completion of the job as you requested on January 15, 19–.
8. Installation of the new transformer at Substation C is complete, and here is my report, as you requested on January 15, 19–.

A practiced writer could devise many other variants with substantially the same content.

If you are writing a periodic report or a progress report submitted as part of a regular routine, omit the "contact" phrase.

The organization of the body of the report will vary according to the subject. If it is more than a page long, however, a brief summary at the beginning is desirable.

To facilitate easy and quick reference to any part of the report, use a separate paragraph for each phase of the subject and begin each paragraph or section with a heading. In its use of headings the short report differs from the ordinary business letter. If you enumerate points of coordinate value, particularly conclusions or recommendations, number and separately paragraph them.

Stop when you have given all the information required. Do not end with superfluous compliments and meaningless statements. It is quite unnecessary to conclude, "I hope this is the information you desire," or "If there is any other information you would like to have, please feel free to call upon me." A reference to future plans, however, may be of interest to the recipient; for example, "A more detailed report will be submitted when the installation is completed."

Ending the report with a participial phrase ("Hoping this meets with your approval, I remain," etc.) is no longer fashionable.

When a letter report runs longer than a page, start succeeding pages not less than an inch from the top. To insure that they will not become separated and lost, you may list at the top identifying data such as name of recipient, page number, and date. The top of the second page would then start as follows:

I. Magnin & Co. page 2 March 1, 19--

The text of the report then resumes three lines below the identifying data (1½ inches from the top). These same data may also be listed in a column, like this:

I. Magnin & Co.
Page 2
March 1, 19--

The second page should contain at least three, preferably five lines. If at first writing you have only a line or so left over from the initial page, rewrite the whole.

COMPLIMENTARY CLOSE

The conventional complimentary close for the short report is "Respectfully submitted," especially when you are reporting to a superior or to a company or organization. Other complimentary closes like "Yours truly" and "Sincerely yours" are appropriate when the relationship is between equals, as, for example, between architect and client.

Place the complimentary close slightly to the right of the middle of the page, two spaces below the body of the letter.

SIGNATURE

Write your signature in longhand just below the complimentary close. Below the penned signature and about four spaces below the complimentary close type your name, the first letter in line vertically with the beginning of the complimentary close.

Below the typed signature give your title or identification. You may need two lines; for example,

Respectfully submitted,
(Penned signature)
John W. Thomas
Director of Research
Engineering Experiment Station

SUPPLEMENT

At the left margin and one or two spaces below the last line of the signature you may type the initials of writer and typist and an abbreviated record of enclosures; for example,

JWT/emc
encl.

Optional arrangements of the parts of the short report are illustrated in the models on pages 188-194.

Military Form of Layout

A variation from the more common forms of the letter report is the modified military form, now sometimes used in business and engineering as well as in the armed forces. The usual heading is typed, but in place of the inside address and salutation three headings are set at the left-hand margin, thus:

FROM: John C. Brown
TO: Joseph E. French, Superintendent
The Waxley Engineering Corporation
5400 South Ardmore Street
New York, New York 10036
SUBJECT: Installation of New Transformer

The salutation and the complimentary close are omitted, but the inside address and the signature are included as in more conventional forms. Side or paragraph headings are used in the body of the report, instead of numbered paragraphs as in military reports.

This modified military form, though more formal than the usual business-letter layout, is direct and convenient and can be safely used for any short report. Note that the FROM, TO, and SUBJECT headings are used in the memorandum report also.

THE MEMORANDUM (OR MEMO) REPORT

The memorandum report is common and useful *within* an organization in situations that do not call for a formal business letter. It may be as short as a letter, or it may extend to several pages. Sometimes a special form, with printed headings for the principal elements, is provided.

The elements in their usual order are these:

1. Report number (not always given)
2. Subject
3. Person or persons to whom the report is addressed
4. Author
5. Date
6. Introduction (sometimes without heading), stating purpose and giving any essential background information
7. Sections of the body, marked by suitable headings, presenting facts and explanations for any conclusions and recommendations
8. Conclusions (if stated)
9. Recommendations (if made)

On pages 188-194 are examples of short reports of different kinds illustrating different layouts.

EXAMPLES OF SHORT REPORTS

The five short reports that follow are of different types and illustrate different forms of layout. The fourth was typed on a printed letterhead, the last on a printed memorandum form.

A. Inspection report; semiblock form, with paragraphs indented; *Attention* line; sideheads underlined, and in separate lines, unpunctuated; heading on second page
B. Progress report; modified military form; *Date, To, From,* and *Subject* headings; no salutation; indented paragraphs; underlined paragraph headings followed by period and dash; complimentary close

C. Examination report; modified military form; *From, To,* and *Subject* headings; indented paragraphs, with headings followed by period and dash; supplementary reference to data sheets, not included
D. Preliminary report; printed letterhead; pure block form; *Attention* line; paragraph headings followed by period and dash
E. Memorandum; printed form; *Subject, To, From,* and *Date* headings; sideheads at the margin, underlined; indented paragraphs; no salutation or complimentary close

EXERCISES

As directed by your instructor, write a short report of one of the following types:

1. A *periodic report* on the monthly, term, semester, or yearly activities of an organized group (fraternity, professional society, forum, dormitory, editorial board, etc.).

2. A *progress* report on a piece of research, an experiment, a term paper, a student project, etc.

3. An *examination report*:

   (a) based on inspection of the condition of a building, of the roads and walks on the campus, of a heating system, of the lighting of classrooms, of the conditions for study in the library, of the condition of a machine, etc.
   (b) based on information derived from a study of usage in some scientific or technical journal (limited to one of the points of style covered in the chapter on Mechanical Aspects of Style).
   (c) based on facts derived from a questionnaire concerning some aspect of students' conduct, of people's beliefs, of teachers' attitudes toward the grading system, etc.

4. A *recommendation report* proposing changes of procedure in student government, changes of methods of conducting the reference room in the library, improvement in the conduct of college social functions, revision of the subject matter of a course or the methods of teaching it, etc. (should include factual information on which the recommendations are based).

5. A *research report* presenting the results of experiments or observations, or summarizing the most recently published information on a given subject.

Notes

1. "Form reports," printed sheets with blanks to be filled out by the author, are also a kind of short report. Since they do not require any special skills, they are not included in this analysis.

A.

533 Smithers Drive
Denver, Colorado 80201
March 1, 19__

MacTavish Tool Company
222 Broad Street
Gary, Indiana 46407

Attention: Mr. F. O. Webster, Chief Engineer

Gentlemen:

In response to your inquiry of February 2, 19__, concerning the condition of the gears in the rock crusher we purchased from you last year, I have inspected the 37-inch bull gear and the 14-inch driving pinion.

Procedure Followed

During the 1000-hour major overhaul of the crusher we dismantled the machine, removed the gears and pinion, and examined them carefully.

After cleaning them with a naphtha solution and drying them with a compressed-air jet, we checked them carefully for cracks and for signs of wear.

Observations

One half of one tooth of the pinion had broken and fallen into the case. The rest of the teeth were intact and showed no large cracks or signs of failure. The faces of the teeth were worn to the extent that their original shape was considably altered.

There was evidence that the oil-filler cap had jarred loose at times and allowed dust to fall into the gear box. Inspection of the oil showed that it contained considerable rock dust.

Conclusions

1. Laboratory inspection showed that the broken tooth was the result of a quenching crack which had destroyed at least a fourth of the tooth's load-bearing capacity.

2. The wearing of the teeth had been due to the dust that had fallen into the oil and then gotten between the teeth.

3. Considering the hard use to which they had been subjected, the condition of the gears was satisfactory.

MacTavish Tool Company 2 March 1, 19__

Recommendations

1. I suggest that you provide a better-fitting oil-filler cap and a better seal between cap and gear box.

2. I recommend that you review your foundry procedure to see if you can avoid quenching cracks in your gear teeth.

Respectfully submitted,

(Penned signature)

George Pine
Junior Engineer
Brown Construction Co.

B.

Date: January 16, 19__

To: Miss Jane Smith, Instructor
Department of Food and Nutrition
Iowa State University
Ames, Iowa 50010

From: Susan Brown
Food Science Major

Subject: Progress on the project to substitute fish-meal flour for wheat flour in bread.

Purpose. The purpose of the project is to develop a recipe for bread using high-protein fish-meal flour in combination with wheat flour. The problem is to find the maximum amount of fish-meal flour that can be substituted to make bread with good texture and taste and acceptable volume.

Procedure. The project requires the following steps:

1. Formulating a basic recipe and experimental recipe with varying percentages by weight of substituted fish-meal flour.
2. Selecting and training a taste panel.
3. Preparing and baking test samples.
4. Evaluating the quality of the bread.

Work Completed. Since approval of the project on January 6, 19__, much of the preliminary work has been completed.

1. The basic recipe has been formulated. (See Table I.)

TABLE I

Basic Recipe

Dry yeast	8 grams
Water	306 grams
Brown sugar	67 grams
Salt	6 grams
Hydrogenated fat	33 grams
All-purpose flour	345 grams
Whole-wheat flour	240 grams

Miss Jane Smith 2 January 6, 19__

Experimental Recipes

Fish-meal flour to be substituted for an equal amount of all-purpose and whole-wheat flour in the following proportions:

5 per cent	approximately	30 grams
10 per cent	approximately	60 grams
15 per cent	approximately	90 grams
20 per cent	approximately	120 grams
25 per cent	approximately	150 grams

2. The percentages (from 5 to 25%) of fish-meal flour to be substituted for equal proportions of the total weight of all-purpose and whole-wheat flour have been determined.

3. The test panel of eight members has been randomly selected from Ralston-Purina employees, and they have been oriented to the kinds of evaluation needed. By testing fish-meal flour in sugar cookies they have learned to recognize its characteristic taste and odor.

Work Remaining. The coming week, January 21-25, will be spent making control samples from the basic recipe and test samples with the various proportions of fish-meal flour.

The week of January 28 through February 1 will be spent evaluating the samples. Measurements of height, weight, and firmness are to be made with precise instruments. The taste panel will subjectively evaluate the test samples for taste, texture, and overall acceptability.

Conclusion. The project is moving along on schedule, and the final report should be completed by the time set, February 10.

Respectfully submitted,

(Penned signature)

Susan Brown
Food Science Major

C.

16 Pammel Court
Ames, Iowa 50010
November 17, 19__

From: Raymond R. Landor

To: Professor W. Paul Jones
Department of English
Iowa State University
Ames, Iowa 50010

Subject: Use of the Suspension Hyphen

In compliance with your assignment of November 11 I am submitting this report on the use of the suspension hyphen, as illustrated by the hyphen after "3" in the phrase "about 3- to 50-kw rating."

Field of Study. - The investigation was restricted to the November and December 1964 issues of Factory Management and Maintenance.

Method and Scope of Investigation. - All of the examples containing suspension hyphens except those appearing in advertisements were marked and recorded on a data sheet. Special care was taken not to overlook phrases where suspension hyphens had been omitted.

Explanation of Data Sheet. - Phrases were listed in the order in which they were encountered, without noting the titles and authors of the articles where they were found. In this study only the usage of the publication, not of particular writers, was regarded as significant. Under the first subhead are listed the 46 phrases with suspension hyphens; under the second, the one phrase with no suspension hyphen.

Conclusion. - The investigation showed conclusively that Factory Management and Maintenance in these two issues was consistent in its use of the suspension hyphen. The one example without the hyphen could be attributed to careless proofreading.

Respectfully submitted,

(Penned signature)

Raymond R. Landor
Engineering Student

Two data sheets

D.

RICHARD HOLTON & COMPANY
Consulting Engineers
146 Sixth Street
Ames, Iowa 50010

January 6, 19__

The Charter Manufacturing Company
Newton, Iowa 50703

Attention: Mr. William F. Hart, Superintendent

Gentlemen:

As you requested on January 1, I am submitting this estimate of the increased production of chemical rocks to be expected from the use of six 30-gallon pressure boilers which you have purchased.

Present Production. - The total capacity of the 30 bottom-heated pails you now use is 900 pounds per hour of white, purple, and green rocks, or 600 pounds per hour of red and yellow rocks.

Estimated Production with Pressure Boilers. - With a steam pressure of 69 pounds per square inch the production will be as follows:

1. Individual capacity: 800 pounds per hour of white, purple, and green rocks, or 600 pounds per hour of red and yellow rocks.

2. Total capacity: 4,800 pounds per hour of white, purple, and green rocks, or 3,600 pounds per hour of red and yellow rocks.

Estimated Increase in Production. - The six pressure boilers will make possible an increase in production of 4,200 pounds per hour of red and yellow rocks.

Accuracy of Estimate. - This estimate, based on the steam pressure and the temperature of the heated surfaces of the boilers, may be as much as 10 per cent in error, because little information is available on the evaporation of these chemicals from such containers.

Final Report. - A final report on the capacity of the six pressure boilers will be made as soon as tests can be run.

Respectfully submitted,

(Penned signature)

John L. Glenn
Production Manager

JLG/abc

E.

Memorandum: A-216

BOLTON MANUFACTURING COMPANY
1600 Penn Avenue
Akron, Ohio 44309

SUBJECT: PROGRESS ON INVENTORY

To: J. H. Regan, Purchasing Agent
From: John B. Gould, Foreman
Materials Department
Date: November 4, 19__

As instructed in your memo of October 28, this department has been working on the physical inventory to be completed by the end of the year. Below is a summary of progress at the close of work November 1.

Progress toward Completion of Inventory

Class of Material	Per cent Completed
Class A	100
Class B	100
Class C	85
Class E	30
Class F	63
Class M	78
Average Completed	76

Conclusions

The average of all classes completed does not truly represent the progress of the inventory. Classes A and B, completed, contain almost twice as many items as either Class E, 30 per cent completed, or Class F, 63 per cent completed.

A more accurate method of determining percentage of work completed would be to weight each class according to the number of items in each. Such a method would show that the inventory is approximately 85 per cent complete.

Signed (penned signature)

Approved ______________

16

Long Reports

A report is usually submitted in long-report form when its text is more than four or five typed pages. In appearance and make-up the long report differs from the short report in having a cover, a title page, a letter of transmittal, a table of contents, an introduction (sometimes headed foreword, digest, or abstract), and frequently an appendix or appendixes. Long printed reports may also have lists of tables and illustrations, a glossary of technical terms, and an index.

In other respects than these there is no difference between a long report and a short one. All of the types of reports listed in chapter 15 may be written in either form. The differences are in length, scope, and outward appearance.

THE COVER

The long report usually has a cover; the short report does not.

Most stores selling books and stationery stock a number of different kinds of report covers, varying in color and flexibility and having different kinds of built-in devices to provide for binding, usually at the left-hand edge. The variety of types, colors, and binding devices permits the exercise of some degree of individual taste in deciding upon the external appearance of the report.

Most covers have an embossed space on the front in which can be pasted a label with identifying data—usually the title of the report, the name of the author, and the date; sometimes other facts as well. Enough information should be given that the report can be identified without having to be opened.

Below is an example of such a label for a report cover:

Report on

POSSIBLE LOCATIONS

for

A NEW MUNICIPAL BUILDING

at

AMES, IOWA

Terry M. Galvin
August 13, 19__

It is permissible, often desirable, to make the report distinctive in outward appearance by adding a decorative design, a photograph, or a figure of some kind, often in color. There is really no reason why reports should be as unattractive in outward appearance as most of them are. Whatever decoration or photograph or figure is added should be in good taste and clearly related to the subject matter. A report on a newly designed circuit for an oscillograph might properly have on its cover a wiring diagram of the circuit or a picture of the instrument, but not a picture of a scantily clad bathing beauty.

Many large organizations use specially designed report covers, usually with the company's name and with blank spaces for essential data, and often so distinctive in coloring and decoration that they are immediately recognizable. Employees of a company are expected to use such covers when submitting reports in the course of their regular duties.

USE OF HEADINGS

Use headings and subheadings in the texts of all reports for two reasons:

1. They enable the reader to find quickly the facts wanted. Like textbooks, reports are used for reference, and are not necessarily read through from beginning to end. The absence of headings would lead to irritation and loss of time while your reader groped about trying to find the section wanted.
2. They make your plan clear to the reader, especially when arranged in table-of-contents form. Knowing what the plan is, your reader more readily understands the whole subject and the relations between its parts.

A third reason might be added: the use of headings almost forces you to arrange your material in logical order. If the table of contents is to present any kind of logical map of the subject, you cannot sit down and put your facts on paper in the haphazard order in which they are likely to occur to you. The actual writing of the report becomes much simpler when you plan the sequence of headings and subheadings at the beginning.

(Review the section on the use of headings in chapter 4.)

Do not hesitate to use many headings. If you find any section running more than a page or two in typed form, scrutinize it carefully to see if you cannot break it down into subsections with subheadings.

Make the headings concise—usually words or phrases, not sentences. The grammatical form may vary from one series of headings to another—nouns, gerunds, infinitives, prepositional phrases, etc. Main headings may be nouns, and subheads some other construction; but in any single series of coordinate headings use the same grammatical construction. To fail to do so is to violate the principle of parallelism.

Compose the sentence following a heading so that it is complete in meaning without the heading; avoid pronouns referring to the heading. If necessary, repeat the words. For example, if the heading is "Reversible Process," write "A reversible process is one in which, etc.," not "This is one in which, etc." Grammatically, the heading is not a part of the paragraph, and the pronoun is left without reference.

TYPING OF THE REPORT

The binding device in most report covers conceals a half inch or more of the left edges of all pages. Therefore you must leave the margin on the binding edge correspondingly wider.

Margins

Leave about an inch of margin at the top and bottom and right side of each typed page, and at least a half inch more at the left side to allow for binding. Number the pages at the top.

Spacing

Double-space the text of the report beginning with the introduction. The double-spaced page is easier to read and (important in reports written by students for practice) leaves space for interlinear corrections. Although single-spacing is common in engineering and industry, it is not recommended for the beginner. Use single-spacing only in short reports, proposals and letters of authorization, bibliographies, and tables of contents. When single-spacing, double-space between paragraphs and before and after center headings and marginal headings.

Form and Layout of Headings

A distinctive mark of the text of reports is that headings and subheadings are used frequently. For a detailed explanation see the section in chapter 4.

ELEMENTS OF THE LONG REPORT

Usually the elements of the long report are arranged as follows:

1. Title page
2. Proposal (not always included)
3. Letter of authorization (not always included)
4. Letter of transmittal

5. Table of contents, including lists of figures and tables
6. Introduction (or digest, or abstract)
7. Body
8. Conclusions and recommendations
9. Appendix (not always included)
10. Index (only in long printed reports)

Title Page

Though there is no standardized form, the title page usually consists of four (often only three) parts arranged from top to bottom in the following order:

Title
Name and identity of the recipient (often omitted)
Name and identity of the author
Place and date

On many title pages suitable connecting words are used so that the whole forms a unit: "Report on . . . . Submitted to . . . . By . . . ." On others such words are omitted, as on the title page of a book.

In many organizations specially printed title pages (as well as covers), with blank spaces to be filled in with essential information, are provided. The arrangement and contents of such title pages may vary from those of the form described, and may include other data, such as the number of the report, a space for the approval signature, security and proprietary notations, etc.

Title

The title of the report may be merely the subject, or it may begin with the words "Report on," "A Progress Report on," "The Annual Report of," "A Study of," etc. It indicates exactly and completely what the report is about. The title of the average book or article is short (usually not over four or five words) and is frequently intended to arouse curiosity and stimulate interest rather than to give information. "For Whom the Bell Tolls" and "It Never Can Happen Again" are long titles, but neither gives the slightest clue to the contents of the books. Often one has to read the book before the significance of the title is evident.

Titles of reports, on the other hand, are intended primarily to give information. They are as short as possible without being cryptic or incomplete; but if it is impossible to indicate the subject in a few words, then many words, up to as many as 40 or 50, are employed. Completeness is more important than conciseness. Baker and Howell, in *The Preparation of Reports,* cite these examples as illustrations:

> A Study of Contact Potentials and Photoelectric Properties of Metals in Vacuo; and the Mutual Relation between these Phenomena
>
> Progress Report of a Special Committee Consisting of the Chief Engineer of the Board of Estimate and Apportionment, the Commissioner of Docks, the

Commissioner of Plant and Structures, and the Engineer of the Borough of Richmond Concerning the Negotiations with the Trunk Line Railroad Companies with Respect to the Brooklyn-Richmond Freight and Passenger Tunnel Project and the Elements of Difference between the Narrows Tunnel and Port Authority Plans

The average title, however, is likely to be shorter; for example,

Surface Tension of Molten Metals
Report on the Performance of Propeller Fans
Report on the Composition of California Lemons
Report on Impact in Steel Railway Bridges of Simple Span

When the title is too long to look good in a single line on the title page, give careful attention to the grouping of words. Set the important phrases in separate lines, with the less important connecting words between. Note the grouping in the following titles:

(1)

Report on
The Preliminary Survey
of
The Manufacturing Methods in the Furniture Factories
of
Grand Rapids, Michigan

(2)

A STUDY OF WAGE-PAYMENT SYSTEMS
suitable for

THE CENTRALIA MANUFACTURING COMPANY

The foregoing titles also illustrate acceptable practice in the use of capitals in titles. Note that when you use only initial capitals you do not capitalize articles (*the, an,* and *a*) and connecting words except when they stand at the beginning of the significant phrases; you do not capitalize connecting words between significant phrases.

Name and Identity of Recipient

When you have prepared the report for a particular person or organization, put the name of the recipient below the title, usually preceded by the words "Submitted to." When the person has an official title, add it after the name, as in the following examples:

(1)

Submitted to
R. C. Memora, President

(2)

Submitted to
The Grand Rapids Association
of
Furniture Manufacturers

This item is absent from the title page when the report is intended for the general public and is not addressed to a particular person or organization.

NAME AND IDENTITY OF AUTHOR

The third item on the title page is the name of the author (or authors), usually preceded by the word "By" and followed by a title or identifying phrase; for example,

(1)

By
Ralph D. Smith
Consulting Engineer

(2)

By
Samuel E. Blythe
and
John R. Tunis
Mechanical Engineering Students

PLACE AND DATE

Write the place and date at the bottom of the page in two separate lines; for example,

Grand Rapids, Michigan
March 17, 19--

LAYOUT

Center the elements of the title page vertically, with sufficient spacing to give a well-balanced layout.

To insure correct centering, lay out the title page first in longhand, and then count the number of letters and spaces in each line. If the middle of the page is at 42 on your typewriter scale, locate the beginning of each line on the page by dividing the total number of spaces by 2 and subtracting from 42. For example, the line

A Study of Wage-Payment Systems

has 31 letters and spaces: $42 - 15\frac{1}{2} = 26\frac{1}{2}$. Disregard the fraction and start the line at 26. Since some lines will have an odd number of spaces and some an even number, perfect balance is difficult to attain.

In laying out the title page—as well as all succeeding pages of the report—remember that when a binding device will conceal the left-hand edge of the sheet, you must make that margin correspondingly wider.

The title page should be typed without error. If you make a mistake, discard the sheet and start again. Erasures are likely to detract from the favorable first impression that the title page should make upon the reader. The following page is a typical example of layout.

The Proposal and the Letter of Authorization

As elements of the long report, either the proposal or the Letter of Authorization may be omitted.

Your work on a report may begin with the writing of a proposal (see chapter 17). If you have written one, you may include it with the rest of your report, usually before the letter of authorization (if it is included) and the letter of transmittal. If the proposal contains the preliminary statement of the problem and outlines the procedure to be followed, the approval of your project is likely to have been oral or in the form of an initialed "OK" and no formal authorization was written.

If you have received a letter of authorization, insert it in your final long report, after the proposal (if there was one) and before the letter of transmittal. It tells you what you are expected to do. It should give you a clear statement of the problem you are to solve; it asks the questions you are supposed to answer in the report. Type the words "letter of authorization" at the top of the page. Include all the usual elements of a business letter.

Either the proposal or the letter of authorization—whichever initiates the project—should be complete and detailed in its statement of what your report is expected to contain. If it is not, further correspondence between you and the authorizer or proposer will be necessary. State the problem completely so that no unanswered questions will come up during the work on the project and the preparation of the report.

Letter of Transmittal

Lay out the letter of transmittal as you would a business letter, with all the conventional elements, and usually with the heading "Letter of Transmittal." Address it to the person or organization for whom you have prepared the report and sign it in ink.

As in the business letter, begin the letter of transmittal with the "contact" phrase referring to the order, request, or commission authorizing the report, and then state that you are submitting the report; for example, see page 205.

Layout of a Title Page

Report on
TWO METHODS OF MOUNTING THE CIRCULATOR
of the
DOPPLER RADAR ANTENNA

Submitted to
Mr. Kenneth J. Engholm
Chief Engineer
Avionics Division
Collins Radio Company
Cedar Rapids, Iowa

By
David Ostergaard
Mechanical Design Engineer
Department C, Avionics Division

Collins Radio Company
Cedar Rapids, Iowa
November 10, 19__

In compliance with your request of March 1, 19–, I submit this report of a study of wage-payment systems suitable for the Centralia Manufacturing Company.

This beginning is admittedly rather stereotyped. For suggested possible variations of the trite wording, see page 183 of the chapter on Short Reports.

The remainder of the letter of transmittal is not standardized. In its shortest form it includes only a statement like the one given; but you may use the letter to call attention to special features of the report, or of the investigation upon which the report is based. In the letter you have your only opportunity to be personal. You can write here what you would say if you were submitting the report in person: you may acknowledge help received, call attention to the limitations of your report, note points of interest in the arrangement, emphasize conclusions reached or recommendations made, point to the need for further study, express opinions.

The letter of transmittal is not a part of the report itself; it merely accompanies and serves to introduce the report. Any essential information in the letter you should include also in the text of the report.

Avoid formal and conventional conclusions of the body of the letter, particularly those that may sound insincere. To write "I have enjoyed preparing this report for you" when it was really hard work and not at all fun is a courtesy of doubtful value. Participial phrases like "Hoping this meets with your approval" are no longer common.

Letters of authorization and of transmittal are illustrated on pages 205 and 206.

Table of Contents

In the table of contents, display in outline form the headings and subheadings of the report. It has two principal functions: (1) it enables the reader to find quickly any section of the report; (2) it provides a quick and comprehensive view of the plan of the report. The experienced reader is likely to turn directly from the title page to the table of contents to find out what topics are covered and in what order.

Various arrangements and layouts of tables of contents are in common use, but for typed reports the forms illustrated on pages 208 and 209 are recommended. The use of letters and numbers with chapter and section titles is optional, but ordinarily you do not need them unless there are many sections and subsections. If you use them, punctuate them consistently.

Defer the making of the final draft of the table of contents until you have completed the text of the report. Then make the table merely by extracting the headings from the text and arranging them in outline form. The arrangement and the wording of headings are exactly the same in the table of contents and in the report.

Letter of Authorization

IOWA STATE UNIVERSITY
OF SCIENCE AND TECHNOLOGY
Ames, Iowa 50010

Department of Economics November 25, 19__

Dear Marvin:

As we agreed in our recent talk, your assignment in Economics 499a is to write a report on the trend toward field shelling of corn by Illinois farmers and on its effects on corn marketing.

Since you are a student of Agricultural Business at Iowa State and your father is an Illinois corn farmer, the subject is a suitable one for you.

I do not expect you to go into detail on the mechanical aspects of harvesting machinery, nor on the problems of marketing above the country-elevator level.

Perhaps your main source of information will be first-hand evidence from your observations of harvesting and marketing corn in Illinois and interviews with farmers and farm-machinery experts. Other sources you should consult are Doane Agricultural Service, Inc., St. Louis, Missouri; Illinois Crop Reporting Service, Springfield, Illinois; and the Iowa State University Library.

A good report on this subject should be of real interest to farmers and farm economists.

Sincerely yours,

(Penned signature)

Arthur G. Brail
Professor of Economics

Mr. Marvin B. Martin
4750 Stalker
Helser Hall
Iowa State University
Ames, Iowa 50010

Letter of Transmittal

4750 Stalker
Helser Hall
Iowa State University
Ames, Iowa 50010
February 15, 19___

Dr. Arthur G. Brail
Professor of Economics
East Hall
Iowa State University
Ames, Iowa 50010

Dear Dr. Brail:

Subject: Report on the field shelling of corn in Iowa.

In fulfillment of the requirements of Economics 499A, I am submitting the following report on the trend to field shelling of corn in Illinois, and the effects of field shelling on corn marketing.

The percentage of the Illinois corn crop that is now shelled in the field has risen from 18 percent in 1960 to 53.5 percent in 1965. Larger farms in central and southern Illinois are setting the pace. Because of the trend toward farm consolidation, the technical and farm-management advantages of field shelling, and other facts, I have concluded that nearly all of the Illinois corn crop will be field shelled before long.

I have also concluded that further economic research on corn marketing is needed to facilitate the necessary market adjustment precipitated by field shelling.

Respectfully yours,

(Penned signature)

Marvin B. Martin
Student of Agricultural Business

In tables of contents not running over two or three pages, include every heading of all levels of importance. If there are many headings, it is sometimes permissible to exclude from the table all headings below the first or second level (as in this book); if all are included, the table may be so long and involved as to defeat the purposes for which it is made.

If there are many subtitles in a section, you can arrange them most economically in paragraph form, with the items separated by semicolons. If there are few subtitles, columnar arrangement is better.

List only the first page of any section in the column at the right, not the first and last pages. Keep the right margin of the column of figures even.

If you use Roman numerals with section titles, lay them out so that the periods are in the same vertical plane, thus:

I.
II.
III.
IV.

Use the same alignment if the numerals are unpunctuated.

The use of "leaders," a series of horizontal dots, to connect headings with page numbers is optional. Their purpose is to "lead" the eye across the page to the right number. In a typed table of contents it looks best to double-space them: if single-spaced, they look almost like a solid line. Their value is so slight that they are often omitted entirely.

In checking copy for the final table of contents make sure that the page numbers are right and that the items correspond exactly in wording with the headings of the report. If a heading reads "Nature of the Problem," the item in the table of contents should not read "The Problem" or "Nature of Problem" but "Nature of the Problem."

Do not include titles of figures and tables—unless you add at the end whole sections headed "List of Figures," "List of Tables." Such lists are often helpful for a reader who may want to find a particular figure or table. Their captions, however, are not really headings, which are always followed by some portion of the text.

To insure that your table of contents is right, defer making its final copy until you have completed the typing of the rest of the report. Then go through the pages, copy out all headings and page numbers, and arrange them in proper form.

Tables of contents are illustrated on the following pages.

TABLE OF CONTENTS

PAGE

TABLE OF CONTENTS

Introduction (or Digest)

In the form of the long report here described, the "double" report, the introduction, often called digest, contains enough information that it is unnecessary for a busy person to read beyond it in order to learn the essentials. It is a preparatory or initial explaining of a subject, as in the preface of a book. It is as nontechnical as it can be, since reports are frequently used by people unfamiliar with technical details, and forms a complete report in itself. The body of the report, the main text, covers the same ground but in detail and in technical language if the subject is technical.

To meet these requirements two forms of introduction are possible, described below as *Option 1* and *Option 2.* Of these the first is the more common, but the second is more helpful to the reader and is recommended for the student and for any other writer not restricted by rules or precedents.

Option 1

Write the introduction in a single section without subheads, usually under the heading "Abstract," sometimes "Digest," "Foreword," or "Introduction," and usually—not always—in a single paragraph.

Perhaps the reason for this practice is that the section can easily be copied without change in an abstracting journal, where it is customary to publish abstracts without subheads. The subject matter is essentially the same as indicated under "Option 2": the object of the report, the procedure followed, and the substance of the body of the report, along with any conclusions and recommendations, are summarized. Although such an introduction is more compact, it is more difficult for the reader to analyze than when subheads are used.

Option 2

Write the introduction under the heading "Introduction" or "Digest" and break it into short subsections with appropriate subheadings such as the following:

Object of the Report (Investigation, Study)
Abstract (Epitome)
Method and Scope
Sources of Information
Conclusions
Recommendations

If possible the whole introduction should be confined to not more than two typed double-spaced pages.

First, state the object of the report, or of the investigation, study, experiment, or observations described in the report. Confine this statement to a sentence or two.

Second, write an abstract of the *body* of the report, preferably of the *informative* rather than the descriptive type. (Refer to the chapter on Abstracts.) Too frequently a descriptive abstract of a report will contain less information than the reader can derive from scanning the table of contents. In this section summarize only the essential information remaining after you have written the other sections. In other words, do not repeat under this subhead statements that appear in other sections of the introduction.

Third, describe briefly the method and scope of the report. Again, do not duplicate any statements made in other sections of the introduction. "Scope" has to do with limitations: define them clearly so that your reader will not expect something you had no intention of including.

Fourth, acknowledge briefly any sources of considerable information outside your own experience and observation, such as individuals who have given you information or advice, and printed sources used. Sources of minor significance you can acknowledge in footnotes.

Finally, list as briefly as possible the conclusions and recommendations of your report. Number and paragraph these, and arrange them in the same order as at the end of the body of the report. If the list is long, omit them from the introduction, but state that detailed conclusions and recommendations are given at the end of the report.

Repeat the conclusions and recommendations at the end of the body of the report, but with supporting material, especially with the reasons for them. Do not worry about this duplication. Many readers will not get beyond the introduction, and might not find conclusions and recommendations—perhaps the most important part of your report—if not listed there. Reports are frequently not read from beginning to end; rather, readers select what interests them. To give an important conclusion or recommendation adequate emphasis, it may be stated four times in the report: in the introduction; at the end; in the body of the report where the detailed supporting evidence for it is presented; and in the letter of transmittal.

Body of the Report

The body of the report presents in full the information abstracted in the introduction. It is the second part of the "double report," presenting in detail—and in technical language if the subject is technical—the information that you have accumulated.

The main heading of the body (the second main heading of the report, since the first is always "Introduction" or "Abstract" or some similar word) is always essentially the same as the title of the report, perhaps in shortened form if the title is long.

The first subheading will be "The Problem" or "The Nature of the Problem" or some similar phrase under which will be presented a more detailed introductory explanation than was possible in the introduction itself.

For the organization of the remainder of the body no specific outline can be given. The subjects of reports and the possible ways of arranging material are so varied that hardly any two outlines follow the same general plan. You must consider what arrangement of the facts is most clear and systematic, and then make the logic of that arrangement obvious to the reader. Supporting evidence for all conclusions and recommendations should be included.

Since most reports are basically analyses, you may find it helpful to review the suggestions on organization and use of headings in chapter 4.

Acknowledgment of Sources

In reports, as in all other forms of exposition, it is essential to make proper acknowledgment of debts for information or statements used. Indicate the source by any one of the several acceptable and conventional established methods. For more detailed information consult chapter 18, Documentation.

For usage in writing titles, footnotes, and bibliographies and lists of references, see chapter 19, Mechanical Aspects of Style.

Conclusions and Recommendations

The last section of the body of the report may be headed "Conclusions and Recommendations," with the two kinds of statement listed under two separate subheads, "Conclusions" and "Recommendations." Two separate main headings may be used.

The number and the order of conclusions and recommendations must correspond exactly with their number and order in the introduction. The difference is that in the introduction they are stated as briefly as possible; at the end of the report they are listed again, often with a summary of the reasons or the supporting evidence for them. Many readers may skip from the introduction to the end of the report: they should be able to get a clear notion of the basis for each conclusion and each recommendation without having to read all the intervening discussion.

Be sure to distinguish between conclusions and recommendations; that is the reason for using two main headings or two subheadings. Conclusions are generalizations derived from the study of known data and concern the past or the present; recommendations, though they must be based on past or present knowledge, state that something should be done and point toward the future. For example, you *conclude* that a certain design for a machine has been faulty in the past or is now faulty; you *recommend* that the design described in the report be adopted.

Be sure to number and paragraph separately each conclusion and each recommendation.

Appendix

Add an appendix or appendixes to the report if you have materials pertinent but not essential to the text. Tables, data sheets, field notes, charts, computations, diagrams, maps, photographs, etc. often make interesting and essential supplements to the text, but are of minor importance. In planning your report, one of your principal problems may be to decide what material belongs in the text and what in the appendix.

Give appropriate titles or headings to the sections of the appendix and list them in the table of contents. If you have more than one or two items, it is convenient to use capital-letter symbols with the titles or headings. References can then be made economically by inserting a parenthesis "(Appendix A)" or an explanatory sentence in the text: "The data for this graph are given in Appendix B." In the table of contents under the main heading "Appendix" list each item with the reference letter and the complete title. Somewhere in the body of the report refer to and if necessary explain each item in the appendix.

REMEMBERING THE READER

The sheer length of long reports places special obligations on the writer in order to avoid the length becoming an additional burden for the reader. If you recall your own usual reaction when someone requires you to read a fifteen or twenty page report, you will realize that placing the reader's needs above your own as writer is especially important in long reports. Two ways to help your reader are by using tabular, graphic, and pictorial methods of presentation, and by including internal summaries and fresh forecasts.

Tabular, Graphic, and Pictorial Methods of Presentation

Too many scientific writers tend to neglect using tabular, graphic, and pictorial methods of presentation, apparently on the assumption that the text itself is sufficiently clear. Such clarity, however, may only be achieved at the expense of great mental exertion on the reader's part. To reduce the amount of mental effort your readers must spend, use visual methods of presentation for key points, for difficult points, and for points that can be presented in more detail in visuals (such as graphs and tables) than could readily be accomplished in the text itself. Review the section in chapter 4 on visuals, and consult back issues of *Scientific American* for ideas about how to present your points visually.

Internal Summaries and Fresh Forecasts

Even the most interested reader can all too easily lose track of what the report has presented in its previous pages, or of where the succeeding pages are leading. Your abstract, statement of central idea, and headings and subheadings help to remind the reader what material has been covered and what remains to be considered. In a long report, however, you also need internal summaries and fresh forecasts to help the reader to follow your subject's development. Although you may not be familiar with the phrase, "internal summaries and fresh forecasts," you probably now use them to some extent. An internal summary is a phrase or sentence which reminds the reader about what has already been covered, and a fresh forecast is a miniature introduction which prepares the reader for the next section of the report. The two usually occur together, either at the end of one major section or the beginning of another. Here are a few brief examples.

> Like the other units mentioned above, the CRT may also be considered as computer graphics equipment. Since the CRT was defined earlier, its uses for computer graphics will be discussed in the next section.

> The information on other reported lipases is confusing. Each investigator reports different physical properties. This confusion indicates that research to isolate milk lipase is necessary.
>
> *Isolation of Milk Lipase*
>
> The need for developing a process to isolate milk lipase has two sources: scientific research and economic factors.

> Having defined the terms and established criteria for comparison, comparison of the texts can now proceed.

> Some of the tax proposals discussed above are clearly in the best interests of the general public. The business community, however, has somewhat different needs.

> The faith placed in a weapons system can determine military intangibles such as morale. In order to produce the best weapon possible, these intangibles must be considered in tandem with the military requirements.

> Now that the process of the wing's development has been explained, the wing's structure can be analyzed in more detail.

> The poisoning process was very effective, but it was a very expensive practice and left huge numbers of dead fish on the surface of the water. As a result of the impracticality of the poisoning program, the decrease in productivity of the state's reservoirs, and pressure from the public, Texas Parks and Wildlife began a program of stocking non-native, or exotic, fish species.
>
> *Non-Native Fish Programs*
>
> The goals of the Texas Parks and Wildlife non-native stocking programs are threefold.

Using an internal summary and fresh forecast at each major textual division in your report helps your reader make the transition between divisions smoothly; this also serves to re-acquaint tired, confused, or inattentive readers with the report's larger structure.

DEVICES FOR SYSTEMATIC PROCEDURE

The procedure followed in collecting material for a long report, organizing the material, then writing and revising the text is long and complicated. It should not be left to chance, but should be systematized from the beginning. Refer to the section on the composing process and the method of science in chapter 2.

The Work Sheet

Just as a contractor with a deadline to meet sets up a definite schedule of the steps to be performed and the dates when each step should be completed, so you, when faced with a problem to solve and a report to write, should prepare a "work sheet" listing the things to be done, and assigning tentative dates for the completion of each step. Such a guide will help you to complete the steps of your work in proper sequence. Without it, you may find yourself approaching your deadline with your problem unsolved and a report to write hastily before you are ready.

The following general suggestions may be helpful to you in preparing a work sheet for your specific project:

1. Analyze the problem.
2. Decide upon the steps you must take to solve the problem.
3. Collect the information needed by:
   a. Carrying out experiments.
   b. Making field trips.
   c. Conducting interviews.
   d. Preparing and distributing a questionnaire.
   e. Reading printed materials pertinent to your problem.
4. Keep accurate and detailed records of all information collected.
5. Tabulate and analyze data.
6. Reach conclusions.
7. Decide what recommendations to make.
8. Organize your materials.
   a. Make an outline.
   b. Devise suitable headings and subheadings.
   c. Arrange the headings in logical order.
9. Prepare tabular, graphic, and pictorial materials.
10. Decide what belongs in the appendix.
11. Write first drafts of the sections of the report.
12. Revise first drafts.
13. Write the introduction.
14. Type the final draft of the text of the report.

15. Proofread the final draft.
16. Make the title page and the table of contents.
17. Write the letter of transmittal.
18. Bind the report in a suitable cover.

Allow plenty of time for each of these steps, especially for items 11-18, having to do with the actual writing of the report. Many people tend to rush this part of the procedure.

Final Checklist

While you are preparing the final draft of your report, you will find it profitable to check the following points:

COVER

1. Have you provided an attractive cover well suited to the subject of the report?
2. Do the title and your name appear on the cover?

TITLE PAGE

3. Does the title page make a good first impression?
4. Is the subject fully and clearly indicated by the title?
5. Are the author and the person(s) addressed clearly identified?

LETTER OF AUTHORIZATION AND PROPOSAL

6. Does the letter clearly state the problem and specify in detail what is required in the report?
7. Is the letter in an acceptable form?

LETTER OF TRANSMITTAL

8. Is there adequate reference to the authorization?
9. Have you used the letter effectively as a device for explanation and emphasis?
10. Is the letter in an acceptable form and signed in longhand?

TABLE OF CONTENTS

11. Does the table of contents provide an overall view of the organization of the whole report?
12. Is the table correct in form and layout?
13. Are headings of the same order logically and grammatically parallel?
14. Do the items of the table correspond exactly in wording and arrangement to the headings of the report?
15. Are the page numbers correct?

INTRODUCTION (OR DIGEST)

16. Does the introduction adequately summarize the essentials of the report?
17. Is the introduction as nearly nontechnical as you can make it?
18. If you have inserted subheadings, are they the proper ones for your report?
19. Does the listing of conclusions and recommendations correspond with that at the end of the report?

BODY

20. Does the main heading for the body approximate the title of the report?
21. Have you been consistent in your method of indicating the relative importance of headings?
22. Is the paragraphing logical? Is any paragraph longer than it should be?
23. Have you used graphic aids to supplement language wherever possible?
24. In the body of your report have you explained all figures, tables, etc.?
25. Have you properly numbered all figures and tables?
26. Have you properly acknowledged the use of material from printed sources?
27. Have you brought all conclusions and recommendations together at the end of the body of the report?
28. Is each conclusion and recommendation at the end supported by a summary of the evidence for it in the body of the report?
29. Have you checked your manuscript carefully for the following?
    Spelling
    Division of words at line ends
    Hyphenation of compound adjectives
    Writing of numbers
    Use of abbreviations
    Dangling modifiers and trailing constructions
    Agreement of subject and verb
    Reference of pronouns
    Jargon
    Clarity and conciseness

APPENDIX

30. Have you titled and lettered the parts of the appendix to correspond with the table of contents?
31. Is each part of the appendix referred to and explained in the body of the report?

PROBLEMS FOR LONG REPORTS

The variety of subjects suitable for practice by students learning to write long reports is limited somewhat by their knowledge and experience and by the amount of time available for experiment, observation, and investigation. Since a report is derived mainly from firsthand experience, its author cannot look up a subject in the library and merely summarize what is there. The result would be one kind of "report," but not the kind that students would be expected to present.

If possible, find a problem in your own field of specialization. Your instructors in that field can probably suggest a suitable one. If you aren't yet qualified to undertake a "special problem," the following list of topics used by students may help you to find a suitable subject:

Problems Related to Student Activities:

1. What are the parking problems on your campus, and how may they be solved?
2. What is the status of the intramural athletic program at your college or university, and how can it be improved?
3. How do students in your curriculum apportion their time, and how might they improve their schedules?
4. Analyze the programs of your professional society or fraternity, and suggest improvements.
5. Analyze your fraternity's pledge-training program, and suggest improvements.
6. Should English 414 (or any other course) be a required course in your curriculum?
7. What are the habits of students in watching television (or smoking, or reading, or dating)?
8. What plans for rebuilding or remodeling should your fraternity adopt?

Problems Requiring Some Special Knowledge or Experience

1. What can be done to modernize a grocery store (shop, garage, elevator, residence, etc.), and how much would it cost?
2. What are the faults of the bookkeeping system of a certain organization, and how can it be modernized?
3. Analyze the merits and faults of different makes of some home appliance (ranges, washing machines, lawn mowers, etc.) and recommend the best.

4. Which of several markets in your college town should a young married couple on a limited budget patronize?
5. Should the ABC company decide to locate its new factory in X-town?
6. What are the fire and safety hazards in a fraternity house (dormitory, laboratory, factory, etc.) and how can they be eliminated?
7. Analyze your high-school education and decide what its merits and faults were.
8. Draw up, explain, and justify your plans for a compact and efficient home woodworking shop (metal-working shop, darkroom, etc.).

Problems Requiring Technical Knowledge or Experience

1. Conduct a time and motion study of a certain operation and recommend improvements.
2. What are the causes of the dropping of ears in the corn fields in the fall, and how can it be prevented?
3. How can the design of a particular machine or machine part be improved?
4. What are the comparative merits and faults of glued and nailed joints?
5. What drainage and soil-conservation measures should be adopted for the improvement of a certain farm?
6. What are the merits and faults of two electrical appliances and which would you recommend?
7. What are the advantages and disadvantages of the use of transistors in an oscilloscope?
8. Compare two or more recipes for making popovers, and recommend the best.

EXERCISES

A. *Choose a problem which you can solve by conducting experiments, making observations, or referring to your own firsthand knowledge and experience, with only a minimum use of printed sources.*

B. *Set up a situation (probably imagined) in which someone would authorize you to work on the problem and present your solution in the form of a long report. Write the letter of authorization that the situation would call for.*
OR:

C. *(Alternative to Exercise B) Write a proposal—to someone you might expect would authorize you to proceed with your project—outlining in detail what you want to do to solve the problem you have chosen.*

D. *Set up a program (a "work sheet") listing the things you should do in solving the problem chosen and in writing the report. Assign tentative dates for the completion of each item on the work sheet.*

E. *A week before the final report is due, submit a title page, letter of transmittal, preliminary table of contents (a list of the headings to be used), and first draft of the introduction of the long report. If necessary, revise these elements before including them in the final report.*

F. *While you are working on the final draft of your long report, ask yourself each of the questions in the "Final Checklist," pages 216-217.*

G. *Submit the completed long report.*

H. *Study the form, arrangement, and contents of the following reports written by students. To save space the single-spaced typed form is used.*

SALES FORECASTING

for

OLDSMOBILE DIVISION OF GENERAL MOTORS

Submitted to

Mr. Edward Cole

President

General Motors Corporation

By

Larry Crabb

Forecasting Consultant

Ames, Iowa

June 1, 197_

LETTER OF TRANSMITTAL

232 Welch
Ames, Iowa
June 1, 197_

Mr. Edward Cole, President
General Motors Corporation
Pontiac, Michigan

Subject: Sales Forecasting for Oldsmobile Division

Dear Mr. Cole:

I have completed the study of methods of forecasting sales of new automobiles in the Oldsmobile Division of General Motors Corporation.

The following report describes the methods that can be used in planning schedules for one month in advance in purchasing, employment, and production.

The report covers the determination of the variables, the forecasting methods developed, and the accuracy of each method employed.

After a comparison of several business indicators with Oldsmobile sales, five indicators were chosen as variables. These were used in testing the methods. Measurements and comparisons were made to provide an accurate forecasting equation for immediate use by the Oldsmobile Division.

Respectfully submitted,

(penned signature)

Larry Crabb
Forecasting Consultant

TABLE OF CONTENTS

TABLE OF CONTENTS (continued)

INTRODUCTION

This report gives the results of comparing several methods for forecasting the sales of Oldsmobile automobiles.

PROBLEM AND PURPOSE The problem is the lack of a reliable method of forecasting Oldsmobile automobile sales for short periods. The purpose is to recommend a method of accurately forecasting sales figures that can be used in planning schedules for purchasing, employment, and production.

SCOPE This analysis utilizes past Oldsmobile sales for 30 months and 60 months. By application of the forecasting methods developed, a formula is proposed for estimating sales for one month in advance. To attain accuracy, the forecasts are checked against actual sales for the most recent 6-month period. To limit the number of possible equations, only five variables were selected for study and checked for accuracy by the computer.

PLAN OF PRESENTATION Explanations of various business indicators are used to establish the method of selecting variables and to compare the relationships between short-term and long-term forecasting.

The methods of analysis are explained, and the forecast results for the most recent five periods are compared with actual figures to determine the accuracy of the methods. Finally, conclusions are reached and recommendations are made.

FORECASTING THEORY Forecasting theory is based on the assumption that an extension of past performance, if adjusted correctly, will provide information indicative of future performance.

Any method of forecasting must have a built-in adjusting device that will reflect the general economic trend and its effects. The adjustment may be based on leading indicators of business activity or on past errors in forecasting.

MAJOR CONCLUSIONS AND RECOMMENDATIONS

For accuracy, the conclusions are derived from the computer output of data.

The regression analysis for the 5-year period had the smallest error in predicting sales of recent 6-month periods.

From the evidence derived from comparisons of the measurements for accuracy, it is concluded that the predicting equation

$$\text{Oldsmobile sales} = -2715.563 + 0.0685\ (\text{total car sales})$$

is the most accurate of those tested, and can be used successfully to forecast Oldsmobile sales for one month in advance.

It is recommended that the equation be used by the Oldsmobile Division of General Motors for forecasting sales.

It is also recommended that research be continued with the use of different variables and new methods.

PREFORECASTING ANALYSIS

This analysis lists and describes the various indicators that may be used in equations for estimating future sales.

BUSINESS INDICATORS

Before representative variables are selected and forecasting methods are developed, Oldsmobile sales must be compared to several business indicators, such as competition, industry sales, and gross national product (GNP).

Competition Competition is high in the automotive industry. Style and design are emphasized, as well as the improvement of engineering aspects of cars. Innovations often provide competitive advantages by hastening the obsolescence of earlier models.

> Indicative of the sharp competition in the field, the index of wholesale auto prices compiled by the U. S. Bureau of Labor Statistics in September, 1968, showed that auto prices declined about 2.5%since 1959, compared with approximately a 20% gain for the consumer price index of all industrial commodities. If the auto manufacturer increases the quality content of a car in terms of its safety, performance, durability, or comfort, the Bureau of Labor Statistics deducts the value of the increased quality from any price boost. [1]

[1] All reference numbers in this report are to the sources given in the Reference List on page [omitted]

General Motors has traditionally been strongest in the medium- and high-priced segments of the market, which include Oldsmobile, Pontiac, Buick, Chevrolet, and Cadillac. Oldsmobile does not compete in the compact market, the smallest Oldsmobile being the intermediate F-85. [2]

Industry Sales The "Big Three," General Motors, Ford, and Chrysler, accounted for all but 9.8% of U. S. automobile sales in 1965. Foreign car sales accounted for gains up to 10.4% in 1956, 12.2% in 1967, and 13.3% in 1968. In that year General Motors produced 51.9% of the domestic automobile output, and Oldsmobile Division produced 13.9% of General Motors' total, or just over 7.2% of the total production. Later figures are not yet available. [3]

Therefore, with nearly 52% of the United States total, General Motors is the world's largest automobile manufacturer, and Oldsmobile is one of the five divisions of General Motors production. [1]

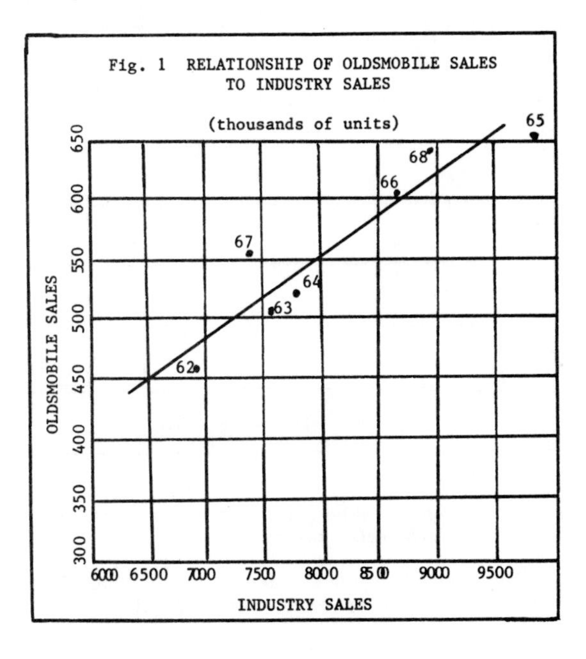

As is evident in Figure 1, Oldsmobile sales are directly in proportion to industry sales. This relationship was very close except in 1967, when the Olds F-85 sales increased while industry production was down.

Gross National Product and Disposable Income Since 1959 General Motors' sales have increased 115%, while the Gross National Product (GNP) increased 88%.

Figure 2 shows the relationship between Oldsmobile sales and GNP from the time the Oldsmobile Toronado was introduced through 1968. During the whole 4-year period no

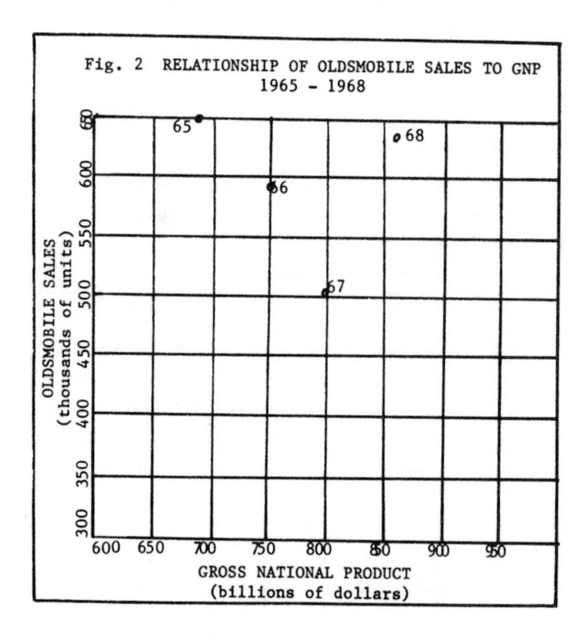

relationship seems to exist; and if 1968 is dropped, there would be an inverse relationship.

Forecasting Oldsmobile sales with only the GNP as a variable would be useless because of the peculiar situation shown in Figure 2.

After the GNP is broken down to disposable income, and it is assumed that Oldsmobile sales follow total passenger car production for the entire period of 1967-1968 as they did in the period of 1962-1968, a comparison can be made to the disposable income. Refer again to Figure 1.

Although the relationship seems somewhat more direct than that to GNP in Figure 2, Figure 3 is also inconsistent when applied to short periods of time. And the short period is esepcially important when the forecast is for one month in advance, as in the methods described in this report.

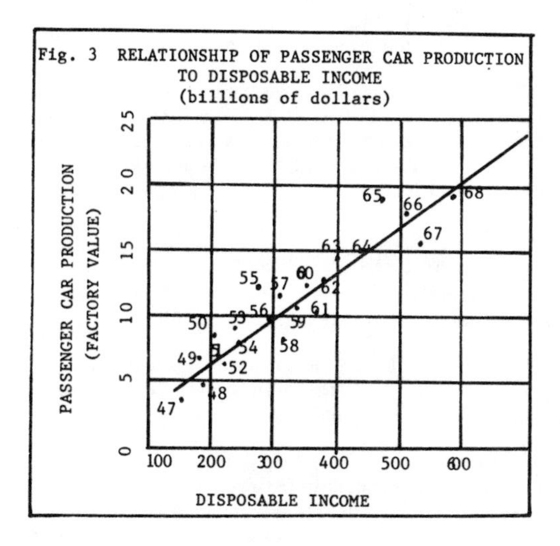

Fig. 3 RELATIONSHIP OF PASSENGER CAR PRODUCTION TO DISPOSABLE INCOME (billions of dollars)

INDEPENDENT VARIABLES From the analysis of indicators that correlate closely with Oldsmobile sales, it now is possible to choose variables for use in forecasting.

Limit Although it is possible to select variables without analysis, the results would be inaccurate without requiring unnecessary computations with the useless variables.

Therefore, to simplify the computations a limit was set to the number of variables. Analysis permits some inconsistent variables to be thrown out before they are used in making forecasts. To hold computer costs down, the limit in the number of variables used was set at five.

Choice The final choice of variables to be used in the trials of forecasting methods is actually a knowledgeable guess. The analysis and general business conditions indicate what variables are significant, but do not justify a final choice of the ones that should be used. The final choice is based on facts, but must be made by a best-guess method. At the start, then, the following five variables were chosen for use in forecasting Oldsmobile sales:

Total personal income
Total automobile sales (industry sales)
Total steel production
Total installment credit
Total magazine advertising

Total industry sales were shown to be directly related to Oldsmobile sales in Figure 1. The other variables chosen were selected in exactly the same way. The final figures indicate which combinations of methods and variables give the most accurate forecast.

FORECASTING METHODS

Two methods of forecasting sales were considered, regression analysis and exponential smoothing.

REGRESSION ANALYSIS Regression analysis is a method of forecasting sales which attempts to estimate values for dependent variables from independent variables.

In forecasting Oldsmobile sales, it is possible to find a mathematical relationship between past sales and other business indicators, such as advertising costs, steel production, disposable income, automobile sales, installment credit, or any other variable. The sales figures can be made dependent on these independent variables when they are equated.

Figures for each of the variables are compiled through regression into past periods. These figures from the past are made equal to one another in balanced equations, and the equations are used to forecast the future. This forecast is then checked against actual Oldsmobile automobile sales, and the degree of accuracy is determined. The equation producing the greatest accuracy can be used to predict future sales (dependent variable) by plugging in the appropriate independent variables. The complete equation developed to forecast Oldsmobile sales is given in Appendix A.

Types Regression analysis was applied to Oldsmobile sales for two time periods, 60 months and 30 months.

Regression analysis for 60 months was used for a long-term correlation between the variables. The forecast indicates the correlation over the period specified.

The same method was applied to a 30-month period, and the resulting forecast reflected the more recent correlation.

Accuracy The measurement of error in each monthly forecast is the simple percentage-of-error measurement, where the difference between the forecasted results and the actual results is divided by the actual sales. (Standard error is commonly called standard deviation.)

Finding the coefficient of correlation is another way to measure accuracy in finding the method that produces the least error. For the equation see Appendix B.

EXPONENTIAL SMOOTHING Unlike the regression method, exponential smoothing does not use a business indicator for a variable. Exponential smoothing relies entirely on past Oldsmobile sales and a smoothing constant, known as alpha. The forecast for the sales of each period is based on the forecast for the preceding period, plus a percentage of the error of the forecast for that period. The percentage of error is determined from a table developed for this purpose (Appendix C).

Degree of Smoothing Smoothing is the method of building market trends into the forecast. In regression analysis these trends are evident in the business indicators; but in exponential smoothing, no indicators are used. Mathematical formulas were developed, therefore, to utilize these trends.

Single smoothing, the first order, carries no adjustment for trend. If Oldsmobile sales were at a constant level, this order of smoothing would prove accurate.

Double smoothing, the second order, assumes that the trend will level out. To be accurate by this method Oldsmobile sales would start in one direction and then stabilize.

Triple smoothing, the third order, assumes a fluctuating trend in Oldsmobile sales. The accuracy of the methods in the actual forecasts indicates which order of smoothing most closely fits the Oldsmobile sales pattern.

Accuracy The measures of accuracy in exponential smoothing are identical to those in regression analysis. Therefore, the accuracy of the various methods can be compared to determine which method is most accurate, or which uses the best combination of variables.

FORECASTING RESULTS

The results of experimenting with the two methods of forecasting are summarized in the following paragraphs.

REGRESSION ANALYSIS

The regression analysis formula applied to periods of 60 months and 30 months gave the following results:

Sixty Months The regression analysis of Oldsmobile sales for 60 months from 1964 to 1968 resulted in a correlation coefficient of 0. 8879 for the fifth variable, total car sales. See Table I. The other four variables, in all possible combinations, were automatically rejected because of low accuracy. The resulting predicting equation is:

$$Y = -2715.1563 + 0.0685 X_5 \quad (1)$$

where: Y equals Oldsmobile sales for the next period
−2715. 1563 is a constant
X_5 equals total car sales

TABLE I. RESULTS OF REGRESSION ANALYSIS—60-MONTH ANALYSIS

PREDICTING VARIABLE TOTAL CAR SALES	PREDICTION	ACTUAL SALES	ERROR
657,622	42,331.89	46,495	−8.9%
607,533	38,900.89	41,752	−6.8%
681,218	43,948.21	48,422	−9.4%
876,020	57,292.14	58,927	−2.7%
889,149	58,191.48	58,565	−0.6%
841,878	54,953.42	53,582	−1.1%
STANDARD ERROR	401.8562		

Thirty Months The application to the shorter period resulted in a correlation of 0.87539, when the second variable, magazine advertising, and the fifth variable, total car sales, were used. The other variables were rejected by accuracy checks. The resulting equation is:

$$Y = -2552.418 + 425,649 X_2 + 0.0629 X_5^2 \quad (2)$$

where: Y = Oldsmobile sales for the next period
- −2552.418 is a constant
- 425,649 is a constant
- X_2 equals magazine advertising
- X_5 equals total car sales

TABLE II. RESULTS OF REGRESSION ANALYSIS—30-MONTH PERIOD

PREDICTING VARIABLES				
ADVERTISING	TOTAL CAR SALES	PREDICTION	ACTUAL SALES	ERROR
6.8	657,622	39,167.47	46,495	16%
8.7	607,533	36,092.80	41,752	13.5%
11.3	681,218	40,845.72	48,422	18.5%
11.3	876,020	53,118.25	58,927	9.8%
11.2	889,149	53,941.12	58,568	7.9%
9.3	841,878	50,882.10	55,582	8.8%

STANDARD ERROR 2222.6020

EXPERIMENTAL SMOOTHING The exponential smoothing method produced a trend of the third order of smoothing, and an alpha of 0.05. See Table III. The forecast is based on the following formula:

$$\text{FORECAST} = \text{PREVIOUS FORECAST} + \text{ALPHA}\ (\text{SALES} - \text{PREVIOUS FORECAST}) \quad (3)$$

The complete equations necessary for forecasting by this method are given in Appendix C.

TABLE III. RESULTS OF EXPONENTIAL SMOOTHING

(units) ACTUAL DEMAND	SINGLE-SMOOTH	DOUBLE-SMOOTH	TRIPLE-SMOOTH
55,000	49,267	47,090	46,678
46,495	49,554	52,018	52,018
41,752	49,401	46,784	47,137
48,422	49,019	51,254	39,432
58,927	48,989	46,724	46,227
58,568	49,468	52,212	37,941
55,582	49,923	47,634	44,865

AVERAGE ERROR 8,217

STANDARD ERROR 12,389.9

EVALUATION

The regressive analysis for the 60-month period provided the greatest accuracy in predicting sales for a 6-month period, as was evident in the very small average error of each forecast, and the standard error of 401. The other methods produced standard-error measurements of 2,222 and 12,389.9, too great to make them useful in forecasting. The correlation coefficient of the 60-month regression analysis indicates a close correlation between the variable, total car sales, and Oldsmobile sales. The coefficient of 0.8879 is quite high in comparison to 1.00 as a perfect correlation.

CONCLUSIONS AND RECOMMENDATIONS

From the evidence obtained by comparisons of the measurements of accuracy, it is concluded that Equation 1

$$\text{OLDSMOBILE SALES} = -2715.563 + 0.0685 \text{ (total car sales)}$$

is the most accurate and can be used successfully to forecast Oldsmobile sales one month in advance.

It is recommended that the equation be used by the Oldsmobile Division of General Motors for forecasting sales.

It is also recommended that research be continued with different variables and new methods.

* * * * * * * * * * *

[To save space, Appendixes and Reference List are not reproduced. See Table of Contents.]

EFFECTS OF LIME STABILIZATION

ON SUBGRADE SOIL

by

Daniel T. Crawford

Student of Civil Engineering/Soils

Iowa State University

October, 197_

for

Mr. John Doe, Engineer

Iowa Highway Commission

Ames, Iowa

713 Pammel Court

Ames, Iowa

LETTER OF TRANSMITTAL

713 Pammel Court
Ames, Iowa 55510
October 31, 197_

Mr. John Doe, Engineer
Iowa Highway Commission
Ames, Iowa 50010

Subject: Effects of Lime Stabilization on Subgrade Soils

Dear Sir:

In accordance with your request, I am submitting the following report concerning the effects of lime stabilization on subgrade soils.

In road construction, lime stabilization is a widely used method. But when a soil is stabilized with lime, problems are frequently encountered. I have studied the most common of these problems and have proposed solutions of some of them. By controlling the amount of lime and the moisture content of the soils, the Commission can be sure that lime stabilization will be effective.

Respectfully submitted,

(penned signature)

Daniel T. Crawford
Student of Engineering/ Soils
Iowa State University

TABLE OF CONTENTS

TABLE OF CONTENTS (continued)

ABSTRACT

Effects of Lime Stabilization on Subgrade Soils

Daniel T. Crawford

Ames, Iowa, 197_

The purpose of the investigation was to study the problems arising when a soil is stabilized with lime. Improper control of the process often causes undesirable results. One type of common Iowa soil and three main factors of stabilization were considered: the quantity of lime to add to a soil, the moisture content of the soil, and the effects of delay before compaction. The method used to determine the quantity of lime to add to a subgrade soil was the subject of experiment. Data collected by others were used to determine the moisture content and the effects of delay time on subgrade soil. For the soil sample studied, the moisture content was found to be from 12% to 13%, the quantity of lime was found to be 3% to 5%, and the effect of too long delay was found to be detrimental. Because of the variation in effectiveness on different kinds of soils, lime treatment of only one kind was considered. The conclusions are that the moisture content of the glacial clay found in Iowa must be carefully controlled. The quantity of lime should not exceed 3.5%; if it does, the effectiveness of the additional lime is wasted. Delay between mixing the lime and compacting the mixture should be kept to a minimum. It is recommended that the tests of the soils used in this experiment be run on every soil before lime stabilization is attempted.

INTRODUCTION

This report concerns the effects of lime stabilization on subgrade soils.

Problem

When soil is stabilized by the addition of lime, the desired results are seldom attained. Lime stabilization is often not effective because the many variables are not properly controlled.

Purpose

The purpose of this study was to determine: (1) the proper quantity of lime to be used when a subgrade soil is treated, (2) the correct moisture content of the soil when it is treated, (3) the proper density of the soil at which maximum bearing strength is maintained, and (4) the effects of varying the time between mixing the lime and compacting the mixture.

Scope

Because of the wide range of effects of lime treatment on different kinds of soil, only soils common in Iowa were used. The study was confined to the mechanism of soil stabilization; the effects of using different types of lime were not considered. The ingredients of the lime used were the same as in any lime available for engineering in Iowa.

Plan of Presentation

First, a study was made to determine the proper amount of lime to add to a soil to stabilize it. Because of the large amount of research involved in finding an optimum moisture content for maximum soil strength and maximum soil density, data compiled by others were used to determine these factors.[1] Also data compiled previously were used to determine the effects of delay on the compaction of lime-soil mixtures.[2]

---

[1] George Luke Pitre, Jr., Moisture-Density and Moisture-Strength Relationships of Common Treated Soils, Iowa State University, 1961.

[2] Carl Arnbal, Effects of Delay in Compaction on Compressive Strength of Soil-Lime-Cement Mixtures, Iowa State University, 1961.

Conclusions and Recommendations

Maximum strength of a soil-lime mixture is attained at a moisture content different from that at which maximum density is attained. Compaction of soil-lime mixtures should proceed as soon after mixing as possible; otherwise, losses in strength and density will result. The optimum amount of lime required for soil stabilization is from 2% to 5% by weight.

On a stabilization project of significant size, it is recommended that the tests performed in this study be run on all soils before lime stabilization is attempted. The results of these tests insure proper field control and effectiveness of lime stabilization.[3]

MATERIALS

Three materials were used in the soil samples: soils, lime, and water.

Soils

The soils used in the investigation varied texturally from loam to silty sand, with the same clay materials present in each sample. Two of the soils were glacial clays from central Iowa; the third was a friable loess from southwest Iowa.

To determine the proper amount of lime to add to a soil sample, a glacial clay from near Fort Dodge, Iowa, was used.

To determine the proper moisture content and proper density, a highly plastic clay common in Iowa was used.[4]

To determine the effects of delay time on lime-soil mixtures, a friable loess from samples found along the western border of Iowa was used.[5]

The physical properties of the soils used are given in Appendix C.

---

[3]The technical and semitechnical terms used in this report are listed and explained in Appendix B.

[4]Pitre, op. cit., page 13.

[5]Arnbal, op. cit., page 8.

Lime

Hydrated lime from the U. S. Gypsum plant in Fort Dodge, Iowa, was used in the tests and stored in a metal airtight container.

Water

Distilled water was used in all tests that required the use of water.

METHODS

Four methods for securing the data necessary for reaching conclusions were utilized: determining the proper amount of lime, determining optimum moisture content for maximum soil density, determining optimum moisture content for maximum soil strength, and determining effects of delay time on compaction of lime-soil mixture.

Determining Proper Amount of Lime

The glacial clay from Fort Dodge, Iowa, was removed from the sample container, dried at room temperature, and broken down into its primary particles. Six representative samples of the primary particles were taken, each weighing 100 grams. To each of the six samples was added a different amount of hydrated lime. The amounts varied from 1% to 6% by weight. After the lime was thoroughly mixed with each soil sample, a physical analysis to determine liquid and plastic limits was run on each sample. The procedure followed in the tests were A. A. S. H. O. (American Association of State Highway Officials), Standard Methods of Sampling and Testing, designations T-89-68 and T-90-70. After the tests were completed, the results for each sample were plotted as the ordinates, and the percentages of lime were plotted as the abscissas, resulting in two curves, which were used to determine the proper amount of lime to add to a soil sample. See Figure 1.

Determining Optimum Moisture Content for Maximum Soil Density[6]

Five portions of the soil with the proper amount of lime added were prepared with five different moisture contents, varying from dry to very wet. Each of the five samples was

[6]Pitre, op. cit., page 18.

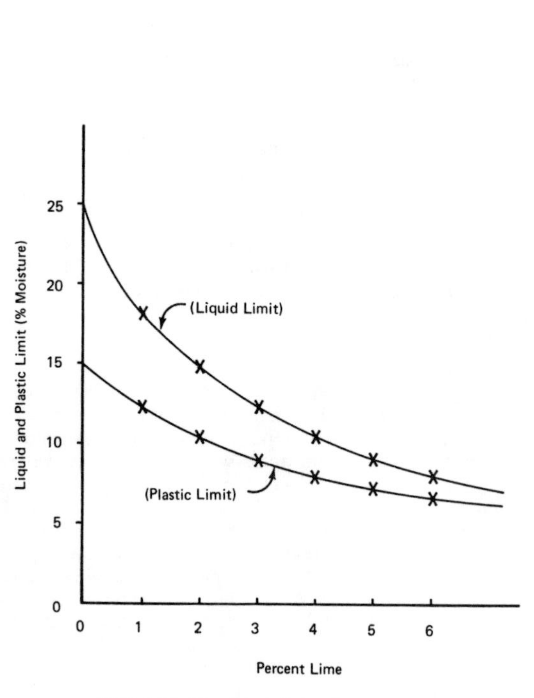

Figure 1. Liquid and plastic limit vs % lime in soil sample.

then compacted in a small mold. The same compactive pressure was used on each sample. From the weight of each compacted sample and from calculations of the moisture content of the soil-lime mixtures, a moisture-density relationship was determined. Equations 1, 2, and 3 below were used to calculate respectively the wet density, moisture content, and dry density. See Appendix D, Sample Calculations.

$$\alpha \text{ wet} = \frac{W_w}{V_m} \qquad (1)$$

where: α wet is the wet density in pounds per cubic foot
W_w is the wet weight of soil in the mold in pounds
V_m is the volume of the mold in cubic feet

$$w = \frac{Wt_w}{Wt_s} \times 100 \qquad (2)$$

where: w is the moisture content of the soil-lime mixture
Wt_s is the weight of dry solids in the soil-lime mixture
Wt_w is the weight of the water in the soil-lime mixture

$$\alpha \text{ dry} = \frac{\alpha \text{ wet}}{1 - w/100} \qquad (3)$$

where: α dry is the dry density in pounds per cubic foot
α wet is the wet density in pounds per cubic foot

The relationship between dry density and moisture content is plotted in Figure 2. The dry density is the ordinate and the moisture content is the abscissa.

Determining Optimum Moisture Content for Maximum Soil Strength[7]

Nine soil samples were prepared, each containing the same amount of lime but a different amount of moisture. Equal amounts of compactive effort were used to pack the samples in molds. The cylindrical specimens were then placed in a compressive strength-testing machine and were

[7] Arnbal, op. cit., pages 12-15.

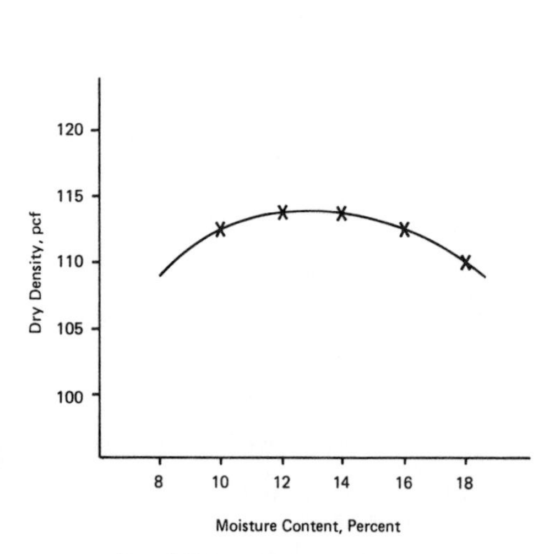

Figure 2. Moisture content vs dry density for lime-treated soil.

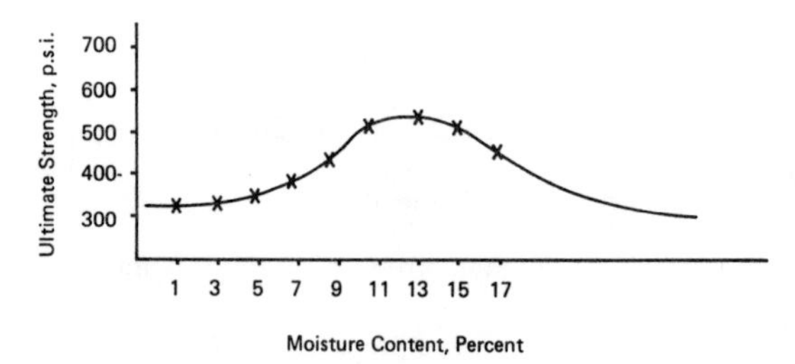

Figure 3. Ultimate strength vs moisture content for a constant amount of lime.

loaded until failure occurred. The strength at failure was recorded as the ultimate strength of the soil specimen. The results of the tests are plotted in Figure 3, with the moisture contents of the samples as the abscissa and the ultimate strength as the ordinate.

Determining Effects of Delay Time on Compaction of Lime-Soil Mixture[8]

Three soil-lime mixtures were prepared, each with the same amount of lime. The mixtures were then placed in metal containers before being compacted. One sample was compacted after 2 hours, another after 6 hours, and the third after 24 hours. Equal compactive effort was used on all three samples. After being cured for equal time-periods, the soil-lime mixtures were placed in a compressive strength-testing machine and subjected to a compressive load until failure occurred. The load at failure was recorded as the ultimate strength of the sample. Figure 4 shows the ultimate compressive strength of the samples as the ordinate, and the delay time before compaction as the abscissa.

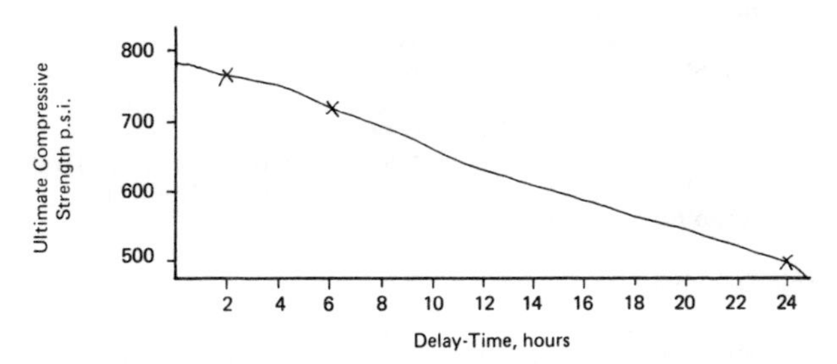

Figure 4. Ultimate Compressive Strength vs Time of Delay in Compaction.

RESULTS AND EVALUATION

Results of the tests are analyzed in the following sections.

Correct Amount of Lime to Add to a Subgrade Soil

For obvious reasons, different amounts of lime are needed to stabilize different types of soil. Experimental results show that a glacial clay which can be found in Iowa requires between 2.5% and 3.5% of lime to be sufficiently

[8] Ibid., page 27.

stabilized. The plasticity of a soil sample is affected by the size of individual grains or particles. Adding lime to a glacial clay flocculates two or more soil particles together. The result is that the new conglomerate of soil grains exhibits properties of a single larger soil grain. A large-grained soil, such as sand or gravel, is more stable than a finer-grained soil, such as a clay or silt.[9]

A non-plastic soil is much like sand or gravel. As the lime content in a glacial clay is increased, the plasticity is decreased. For economic reasons the amount of lime that can be feasibly added to a glacial clay is limited. In the samples tested, the most marked decrease in plasticity occurred with the addition of approximately 3% of lime.

Desired Moisture Content for Lime-Treated Soil

The desired moisture content of a glacial clay found in Iowa varies with the dry density required. If the maximum density is desired, as it often is, the moisture content should be kept at the same level.[10] By experimentation with the glacial clay sampled sampled near Fort Dodge, Iowa, an optimum moisture content of approximately 13% was indicated. The moisture in the lime-soil mixture serves two purposes: first, to react with the lime and clay minerals to bond two or more particles together; and second, to lubricate the conglomerate of soil particles so that they may move more freely and closely together, and thus increase the density of the soil-lime mixture.[11]

Desired Moisture Content for Maximum Soil Strength

Moisture is important to increase the strength of a lime-treated soil. Too little moisture is insufficient to react properly with the soil grains. Too much moisture causes

---

[9]Manual Matoes, Physical and Mineralogical Factors in Stabilization of Iowa Soils with Lime and Fly-Ash, Iowa State University, 1961.

[10]Ibid., page 30.

[11]Melvyn Douglas Remus, Strength of Soil-Lime Mixtures at Standard and Modified A. A. S. H. O. Densities, Iowa State University, 1960.

excessive lubrication between soil grains, and decreases the strength of the bond between the particles.[12] As illustrated by Figure 3, a moisture content for maximum soil strength for the glacial clay of central Iowa is approximately 12%. This figure was arrived at by experimentation with a mixture of glacial clay and lime at various moisture contents.

Effects of Delay on Compaction of Lime-Soil Mixture

Delay-time is the time between mixing lime with the soil and compacting the soil-lime mixture. Hardening of soil conglomerates occurs if the delay is too long. Consequently, when the combined soil aggregations are compacted, small voids are left between the particle conglomerates. The small voids between particles mean lower density for the compacted mix.[13] Any delay between mixing and compaction of the mixture results in a loss in soil density and a corresponding loss in soil strength. From the experiments with glacial clay from central Iowa, the above conclusion is reinforced. The experiments are the basis for a recommendation to minimize delay between mixing and compacting a soil-lime mixture.

CONCLUSIONS

The following conclusions have been reached:

Lime Content

For a glacial clay common in central Iowa a lime content of 3.0% is sufficient for stabilization.

Moisture Content

The moisture content of the lime-soil mixture should be from 12% to 13% to insure a combination of maximum density and near maximum strength.

Delay-Time

The delay-time between mixing the lime with the soil to be treated should be as small as possible.

---

[12]George B. Sowers and George F. Sowers, Introductory Soil Mechanics and Foundations, third edition, Macmillan, 1970.

[13]Arnbal, op. cit., page 44.

RECOMMENDATIONS

The following recommendations are made:

Compaction

For best results, it is recommended that compaction of the lime-soil combination proceed immediately after the lime has been added to the soil.

Continuation of the Tests

On a stabilization project of significant size, it is recommended that the tests described in this report be run on all soils before stabilization is attempted. The results of these tests insure proper field control and effectiveness of lime stabilization.

APPENDIXES

APPENDIX A. BIBLIOGRAPHY

1. Archibold, James, "The Challenge of Change," Boston Society of Civil Engineers, Vol. 60, No. 2, June 1973, p. 33.

2. Arnbal, Carl Anton, Effects of Delay in Compaction on Compressive Strength of Soil Cement and Soil-Lime-Cement Mixtures, Iowa State University, 1965.

3. Clark, Paul, "The Highway Engineer on Trial," Civil Engineering, Vol. 43, No. 9, Sept. 1973, pp. 60-63.

4. Mateos, Manual, Physical and Mineralogical Factors in Stabilization of Iowa Soils with Lime and Fly-Ash, Iowa State University, 1961.

5. Pitre, George Luke Jr., Moisture-Density and Moisture-Strength Relationships of Cement-Treated Soils, Iowa State University, 1961.

6. Reid, Frank, "Secondary Roads Still Rate Second," Better Roads, Vol. 43, No. 8, Aug. 1973, pp. 15-18.

7. Remus, Melvyn Douglas, Strengths of Soil-Lime Mixtures at Standard and Modified A.A.S.H.O. Densities, Iowa State University, 1960.

8. Smartt, Vaugh, "The Interstate Expedition," Constructor, Vol. LV, No. 8, Oct. 1973, pp. 18-25.

9. Smith, Lorraine, "Rockfill Highway Job Puts Earth Movers to Test," Construction Methods and Equipment, Vol. 55, No. 10, Oct. 1973, pp. 61-64.

10. Sowers, George B., and George F. Sowers, Introductory Soil Mechanics and Foundations, Macmillan, 1970.

11. Standard Specifications for Highway Materials, Part II, Methods of Sampling and Testing, 10th ed., American Association of State Highway Officials, 1970.

12. Wong, H. Y., "Engineering Problems in Unsaturated Soils," Civil Engineering and Public Works, Vol. 68, No. 806, Sept. 1973, pp. 759-799.

APPENDIX B. GLOSSARY OF TECHNICAL TERMS

ABSCISSA — The horizontal scale of a cartesian coordinate system

CONGLOMERATE — The aggregation of a few soil particles into one larger soil particle

DELAY-TIME — The time between mixing lime with soil and compacting the mixture

DENSITY — The mass of soil present in a certain volume, expressed as pounds per square foot (psf)

GLACIAL CLAY — A soil consisting of particles less than 5 microns in diameter, indicating the glacial origin of the soil

LIME STABILIZATION — The increase of the positive characteristics of a subgrade soil by change in the properties of the soil elements, principally the clay minerals and the absorbed water

MOISTURE CONTENT — The relative wetness or dryness of a soil; the weight of moisture in a soil sample divided by the weight of dry solids in a sample

OPTIMUM MOISTURE CONTENT — The best moisture content to attain another property desired, such as maximum density

ORDINATE — The vertical scale of a cartesian coordinate system

SOIL STABILIZATION — The change in the physical and chemical properties of a soil to increase its load-carrying capacity

SOIL STRENGTH — The load-carrying capacity of a soil

APPENDIX C

TABLE I. PHYSICAL PROPERTIES OF SOILS

Soil Sample	Glacial Clay	Highly[14] Plastic Clay	Friable[15] Loess
Location of Sample	Fort Dodge, Iowa	Belmond, Iowa	Logan, Iowa
Liquid Limit	25.0	51.0	30.8
Plastic Limit	15.0	29.0	24.6
Plasticity Index	10.0	22.0	6.2
Standard Procter Density (pcf)	110.0	95.0	105.0
Optimum Moisture Content (%)	15.0	23.0	19.0
Specific Gravity	2.64	2.66	N.A.

[14]Pitre, op. cit., page 14.

[15]Arnbal, op. cit., page 9.

APPENDIX D

SAMPLE CALCULATIONS

(not included)

17

Proposals

Proposals are a difficult type of writing to classify. Since they have to do mainly with the future rather than the past or the present, they are not really reports or scientific papers as those types are ordinarily defined. Only recently have the textbooks on technical writing of reports begun to offer any information about them. Yet they are an essential part of the total procedure in getting work accomplished in engineering, science, and business.

DEFINITION

As the name indicates, a proposal is a written offer that you make for someone's consideration—you or a group that you represent offers to do something for somebody else. It is a statement of your desire to perform a piece of work, do some research, or solve a problem, sometimes technical, sometimes nontechnical. It usually explains what the project is, outlines the procedure to be followed, and estimates what personnel, equipment, and money will be required.

As the author of a proposal you will usually expect to profit in some way from the project. Your writing, then, comes close to being persuasive rather than scientific—as those terms are defined in chapter 1 of this book. You will naturally want to present your plan so that the recipient—individual or group—will authorize you to proceed. You may be tempted to magnify the advantages to be derived, or to minimize the problems to be encountered. At the best, however, your proposal can be an impersonal, unprejudiced analysis of the need and the requirements of the undertaking. If it is, it can properly be considered a scientific paper or report.

You will make your proposal to someone with power to authorize the project: your employer, a professor in charge, your superior in the company you work for, a client. If you are a student, you can make it to an instructor, to a committee under whose direction you are working for a degree, or to some individual or company that you hope will sponsor the project.

COMPARISON OF THE PROPOSAL, THE LETTER OF AUTHORIZATION, AND THE REPORT

A proposal is similar in some ways to a letter of authorization. Both or either may be a part of the total report-writing situation. The letter of authorization may request or order someone to initiate or carry out a project, or it may merely request a proposal. That is, the letter might precede the proposal and authorize or solicit it, or it might follow the proposal and authorize the project itself.

Neither the letter of authorization nor the proposal is entirely essential for the initiation of a project. Their equivalents may be in oral form. Or, if the authorization is detailed enough, no written proposal is needed, since the requirements for the project are stated in the letter. If the project has been proposed beforehand, the authorization may be merely in the form of a signed "OK" on the proposal, signifying that the reader approves the plan.

Simply for the records, however, you should prepare either a detailed proposal that has been marked "Approved" or a detailed letter of authorization. Whether one or both of these are bound with the final report or omitted depends on many things: most readers of the report probably will not need to see them.

EXAMPLES OF PROPOSALS

The following are miscellaneous examples of undertakings which might be initiated by the submitting of a proposal:

1. To carry out a chemical experiment that, if successful, would have value to a certain industry.
2. To conduct experiments that would, if successful, determine the cause and possible cure of some plant disease.
3. To consider whether or not to spend $7,685 for planning and constructing expanded warehouse facilities.
4. To conduct experiments that would determine the causes of the poor color of frozen turkey carcasses and find a solution to the problem.
5. To prepare a training manual for the apprentice employees of a large company.
6. To design and build a pilot model of a new airplane for the military forces.
7. To study a particular step in a manufacturing process now being inefficiently performed and propose improvements.
8. To study the deterioration in the surface of an asphalt highway and propose remedies.
9. To analyze the bookkeeping methods of an organization and see if computer methods can profitably be adopted.
10. To survey a given area to discover whether mineral deposits of any value are present.
11. To ascertain the existing methods of abstracting scientific papers in some field and suggest changes.
12. To conduct a survey of the parking situation on a university campus and propose improvements.

These examples, picked at random, illustrate the fact that the requirements for carrying out proposals may range from the effort of a single individual working alone and incurring little expense to that of a large industrial organization employing thousands of people and incurring millions of dollars of expense.

CHARACTERISTICS OF A GOOD PROPOSAL

Since the characteristics of a good proposal are much the same as those of a good report, reference to chapter 16 is suggested at this point. Some characteristics, however, need special emphasis.

Consideration for the Reader

Since you hope that your proposal will be favorably received, you should try to impress the reader favorably. You should emphasize the advantages of the project for your reader if it is carried out successfully. In other words, keep in mind the old rule, *remember the reader.*

Fairness of Presentation

You should present your proposal fairly and honestly, neither overemphasizing the advantages nor ignoring possible objections. The prospective sponsor probably knows less about the subject than you do, but the sponsor is entitled to know the advantages and the objections before approving or disapproving. If, for example, you as a member of a committee are proposing to a university administration that students be prohibited from parking cars on the campus, you should list not only the advantages but also the disadvantages of the proposal, and give some account of the way the plan has worked on other campuses.

Avoidance of Obtrusive Sales Talk

The prospective sponsor is likely to be unfavorably affected by sales talk that is too obtrusive. If the advantages of the proposal clearly outweigh the disadvantages, your reader will be favorably impressed, but direct sales talk will arouse suspicion and merely cloud the issues. If you are proposing that the company you work for buy a hundred Fords for its salesmen, the officers of the company will want factual data on first costs, trade-in values, maintenance costs, fuel efficiency, quality of dealers' service, etc., and not flowery talk about the beauty of the new models, the variety of body styles and colors available, and the number of seconds required to reach 55 miles per hour from a standing start.

Completeness of Information

You should include all pertinent information about the project, anything that might contribute to its success or failure. If the proposal is to undertake some work in the future, you can be excused, of course, for not foreseeing every problem. Some projects that seem to promise desirable and profitable results prove impossible to carry out and have to be abandoned. Scientific and industrial progress often depends upon undertaking tasks beyond one's ability to perform. Today maybe the job cannot be done; tomorrow, because more complete knowledge, better equipment, and greater skills are available, maybe it can be done. To insure the best chance to carry out a project, you should present all the facts. If someone else has already tried to carry it out, it would be a costly waste of time, energy, and money not to report that information and run the risk of another failure.

A Logical Plan of Organization

From a poorly organized proposal your reader is likely to get a faulty conception of the project, and therefore not favor an attempt to carry it out. As in writing any composition longer than a paragraph, you will do well to make a plan: outline what you want to say, set up a series of headings that will indicate your plan, both to yourself and to your reader. If you will need two hours to write the proposal—assuming that you collected your material before writing—you probably should spend at least the first thirty minutes planning.

Clearness of Language

It is not enough to plan or to write a first draft of a proposal according to a plan. You should correct and revise the first draft till its meaning is clear. Clarity is a relative quality, of course; even the most skilled of writers fails to communicate all his or her meaning. Your objective should be to write not so your reader can guess at your meaning, but so that the reader cannot fail to understand. Anatole France, a famous French author, recommended seven revisions and thought that an eighth was desirable to insure that the corrections on the seventh had been made.

Most authors, of all kinds of writing as well as of proposals, lack the patience needed. They get their thoughts down on paper in the first words that occur to them, and then trust to luck that the reader will understand. Too often their luck is bad, and the proposal upon which perhaps their professional future may depend is rejected because it has not been understood. It pays to write clearly.

ORGANIZATION OF THE PROPOSAL

For the organization of the body of the proposal no single plan will meet the needs of all writers. Situations will differ in kind, scope, and degree of complexity. One proposal may be completely presented on a single page in the form of a letter or memorandum; another may be many pages long and require the use of the long-report form, with many headings and subheadings.

Regard the following plan, therefore, merely as one example, and do not adopt it slavishly for use in all proposals. Add to, subtract from, or rearrange the items till you have an outline that fits your needs. This plan is suitable for proposals in letter form involving mainly two people, the one making the proposal and the one receiving it.

1. *Personal references*

   If you have had any previous relationship with your correspondent, especially as an employee, refer to that relationship in your first paragraph. Favorable reaction to it may arouse interest in your proposal.

2. *Summary, or abstract*

   State briefly and clearly what it is you want to do. Don't make it necessary to read to the end to make out what you have in mind.

3. *Motivation*

   Show the advantages to be derived from your project. Demonstrate that the problem you hope to solve exists, and that there is a clear need for its solution.

4. *Background*

   Describe the situation that gives rise to the present need. If previous attempts to solve the problem, or a similar one, have been made, describe them and indicate why they were unsuccessful. Explain why your proposal has a better chance of success than earlier attempts.

5. *Qualifications*

   Describe what experience and what special knowledge and skills you have that qualify you to undertake the project. The better your qualifications, the more likely the response to your proposal is to be favorable.

6. *Probable procedure*

   Describe in detail the procedure you plan to use. Give your reasons for each step, and explain how it is essential to the success of the project. Use diagrams and other visual aids to make these steps clear.

7. *Equipment, materials, and facilities available*

   List the equipment, materials, and facilities needed that are already available at little cost, or that could be adapted to your needs.

8. *Equipment, materials, and facilities not available*

   If any or all of the equipment and materials needed are not available, list them. Indicate where they can be obtained and what problems you may have in securing them. If you would need facilities like an office or laboratory space, specify what those needs would be.

9. *Personnel*

   If your project is more than a one-person job, explain what help you would need. Indicate whether you would need skilled or unskilled help, and whether you know where you can get the help required.

10. *Preliminary problems*

    If you would require special apparatus that cannot be purchased, explain what it is and how you would provide it. If you need to consult other persons, tell who they are and what arrangements you would need to make with them.

11. *Estimated costs*

    List the estimated costs of the equipment, materials, and aid that you would require. State whether your own work and time would be items in these costs.

12. *Preliminary and progress reports*

    Indicate what preliminary and progress reports you would submit. A preliminary report when you are ready to begin your project may be desirable; progress reports during the project are usually expected. For the record they should probably be written rather than oral reports.

13. *Faults of the proposal*

    List all of the faults and weaknesses of your proposal. To do this requires that you be completely objective and that you talk beforehand with people who would be competent to criticize your plan. To omit intentionally the listing of faults would make your proposal an example of nonscientific writing. Your reader will be better able to judge the merits of your proposal if you don't overlook stating its weaknesses.

14. *Merits of the proposal*

    Finally you should be explicit in stating the merits of your proposal and listing the profits and advantages that would result from its adoption. You are entitled to emphasize these merits because eventually your listing of them will—if your proposal is a worthy one—result in a favorable decision.

15. *Conclusion*

    You should question seriously whether any conclusion is necessary. Don't end with superfluous statements like "I hope you will consider this proposal carefully" and "I shall expect your decision very soon." If you have written carefully and have fully explained your proposal, your reader is unlikely to ignore it.

Also, avoid participial endings like "Hoping this proposal meets with your approval, I remain, etc." The ending of a proposal is a place for emphasis. Make it meaningful.

Variations upon and modifications of an outline should be fairly easy if you are prepared to carry out the proposal. *How to Prepare Effective Engineering Proposals,* by Emerson Clarke, lists four basic elements:

1. What we propose to do (the activity)
2. How we propose to do it (technical approach)
3. How we plan to do it (the procedure)
4. What we propose to do it with (facilities and capabilities)

The author adds: "A fifth element is, of course, 'how much we propose to charge.' . . ."

These main divisions would probably fit the needs of most writers, although (as Mr. Clarke goes on to explain) the material in each, at least in a proposal of any complexity, would need to be presented under many subheads.[1]

FORM

The form of the proposal will usually depend on its length and its complexity, and on the reader for whom you intend it. If the project is simple and involves only you and a single reader, it is likely to be in the form of a letter or a memorandum. The proposal for a student's project to be completed in a few weeks would normally be in one of these forms. If the project is complex and would involve many people, it is more likely to be presented in long-report form, with cover, title page, table of contents, many pages of text, and many divisions. Intracompany ("intramural") proposals—those intended to be circulated only within an organization or a single division of an organization—especially when they are limited in scope, are likely to be in the form of memoranda. Those made to an outside company or client or division of government ("extramural") are likely to be in long-report form if they are at all complex, in letter form if they are brief.

Even in its briefest form—a letter or memorandum only a page or so long—the body of the proposal should have paragraphs with headings.

USE OF GRAPHIC AIDS

Ideas and facts are often communicated more clearly if the language of the proposal—like that of the report—is supplemented by tables, graphs, and illustrations. No matter what the proposal is about, the chances are good that it will be more interesting and more effective if you use a graphic aid whenever the opportunity offers. Whole books are devoted to the subject, and so many magazines and newspapers have good examples that illustrations hardly seem necessary.

For examples, see *U.S. News and World Report*, *Newsweek*, *Business Week*, *Fortune*, textbooks in economics, engineering, commerce, science, etc., annual reports of corporations, and numerous government publications. For fuller treatment of the use of graphic aids, see the books listed in the bibliography.

EXERCISES

A. *Write a proposal for a project to be undertaken in one of your college courses, and to be completed and reported upon before the end of the term or semester.*

B. *Criticize each of the following students' proposals to undertake projects to be completed within a term or semester in a course like the one in which you are using this textbook. Can you suggest any improvements in either of them?*

Notes

1. For more detailed outlines of particular types of proposals, refer to the books listed under "Proposals" in the bibliography.

Wallace Hall
4319 Errington
Iowa State University
Ames, Iowa 50010
March 19, 1969

Dr. H. A. Spalt
Research Center
Masonite Corp.
St. Charles, Illinois 60174

Subject: Proposal to compare the amounts of swelling when woods are wetted by different liquids.

Dear Dr. Spalt:

From my contact with you this past summer and through the research project I did for Masonite, I have come to realize and appreciate the importance of understanding the chemistry of board formation. To produce a better hardboard, the relations of the fibers and the forming medium must be understood.

Definition of the Problem

My project last summer revealed that certain organic liquids are better than others for forming fiber mats. The reasons for this appeared to be twofold:

1. The surface tension of the liquid determines how closely individual fibers are brought together.
2. The amount of swelling of the fibers determines how well they will adhere to each other.

The surface-tension phenomenon is fairly well understood, but the amount of swelling varies with the swelling agent. Some method of predicting the amount of swelling would be useful in present practical applications and in future research.

Proposed Solution

The amount of heat generated by wetting wood is known to be related to the amount of liquid sorbed. Since that amount is related to the amount of swelling, we would logically expect the heat of wetting to be related to the amount of swelling. I propose to test this hypothesis by comparing the amounts of heat generated by wetting wood in various organic liquids with the amounts of swelling these liquids produce in wood.

Dr. H. A. Spalt 2 March 19, 1969

Description of Proposed Work

The methods I would use in making this comparison would utilize well known laboratory procedures. The method for determining the amount of volumetric swelling is basically the same as the one I used this summer, namely, impregnation of the samples with the liquid and measurement of the volumetric increases. Modified calorimetric techniques, similar to those described by Stamm (Alfred J. Stamm, Wood and Cellulose Techniques, The Ronald Press Company, 1964, page 202) can be used to determine the heat of wetting. The data from the two phases can then be compared and a conclusion reached whether or not the heat of wetting can be used to predict swelling.

Equipment, Facilities, and Personnel Needed

Since the procedures I would use are familiar to you, I shall not describe the equipment needed in detail. Basically I should be using a vacuum pump, a mercury bath, a calorimeter, and a Beckman thermometer calibrated to hundredths of a degree centigrade. Since I would be able to do this research in conjunction with one of my forestry courses, there will be no expense for personnel.

All of the facilities and equipment needed are available in the Forestry Department's laboratory or can be obtained from the University chemical stores.

Schedule

Although the project should be fairly simple, I do have other course work and thus will not be able to submit a report till the middle of May. I would, however, provide you with a progress report about the second week in April. Since the two phases of the project would run independently until the data-analysis stage, they would be run simultaneously.

The only problems that might interrupt this schedule would concern the modification of the calorimetric techniques. I must develop a technique that is suitable for the available equipment.

Advantages

A simple method of predicting swelling would have distinct advantages. Being able to predict the amount of swelling resulting from the heat of wetting would facilitate finding a better forming medium than water.

Dr. H. A. Spalt 3 March 19, 1969

Even if the project did not provide any immediately useful information, it would indicate other avenues of approach to the problem and possibly provide a springboard for further research.

Sincerely yours,

(Penned signature)

Fred Simon, Jr.

120 Lynn Avenue
Ames, Iowa 50010
December 1, 197__

Dr. James Lowrie
Committee for the Institute
on National Affairs
Iowa State University
Ames, Iowa 50010

Subject: Proposal to prepare guidelines for editing the publicity for the Institute on National Affairs

Dear Dr. Lowrie:

As a member of the Committee for the Institute of National Affairs, I would like to propose some guidelines for the design and preparation of all the publicity for the Institute. In the past, posters, brochures, and fliers have been drawn up carelessly and hastily, without much regard for the budget.

The procedure must be improved if we are to retain our GSB allocation of funds for the Institute. More effective publicity will heighten the interest of students, an important consideration if we are to have effective Institutes in the future.

General Goal

Since I shall have the duty of preparing publicity for the Institute brochure this year, I would like to keep complete records of money spent, materials used, and time required to design the brochure. With it as a guide, in the future even an inexperienced person could set up a good budget.

Method

I would develop the plan and set up guidelines for procedure as I work on the brochure. I would include information of the following kinds:

Copy. I would outline the requirements for copy needed in drawing up the brochure. Included would be the necessary information about the speakers participating in the Institute, facts about the local people who are to serve on panels, and reviews of short summaries of films to be used in the meetings.

Type. I would describe the different kinds of type available to the designer of the brochure, and list the factors to be considered in choosing type faces.

Paper. I would list the different kinds of paper available for printing the brochure, and estimate their costs. I would explain the factors to consider in choosing photographs and art work and deciding about different colors of inks and papers. I would give reasons why these may be effective or noneffective, and why good choices justify the extra cost.

Layout. I would state the principles of good layout that must be considered in designing the brochure: what facts to consider in the placement of copy, art, photographs, etc. Also I would describe the procedure used in fitting a number of words in a given space.

Required Resources

The resources necessary to set up such guidelines are minimal. Of importance to me is time, since my design of this brochure is limited by the printing deadlines and the dates the committee has set for the Institute of National Affairs.

I would need an estimate of the budget the committee will set up for publicity, past records of money spent for publicity, and records of expenditures in recent years. Also I would need your cooperation and that of the other members of the committee in determining the schedule of meetings and speakers for the week of the Institute so that I may begin work on the design of the brochure.

To develop this program I would require no money, but I would need help in getting the biographical information needed for the brochure.

Program

At the last committee meeting November 29 the committee still had open spots for speakers on the calendar. So far as possible I would set up the following schedule and try to adhere to it:

Sat.	Dec. 1	Choose the size and style of brochure. Determine the number of pages.
Mon.	Dec. 3	Design the cover for the booklet.
Tues.	Dec. 4	Edit the biographical material available.

Thur.	Dec.	6	Gather and edit facts about speakers for whom biographical data are scarce.
Fri.	Dec.	7	Copy the biographical information and determine type faces and sizes of type to be used and the amount of space needed.
Sat.	Dec.	8	Lay out the copy for the brochure in proper order, leaving space for art-work, pictures, etc. Letter in the headlines. Specify the type sizes of all headlines and text matter.
Mon.	Dec.	10	Have the artists do the artwork.
Wed.	Dec.	12	Have all copy set up and typed ready for the printer. Indicate where each bit of material is to be placed.
Fri.	Dec.	14	Have all material proofread and approved by the committee.
Mon.	Dec.	17	Send the material to the printer. Request 10,000 copies. Allow one week for delivery.
Mon.	Dec.	24	Have the brochures delivered to the office. Plan to draw up a final report during the Christmas break.
Mon.	Jan.	7	Have the brochures ready for distribution when classes reconvene.
Mon.	Jan.	13-to	Fri. Jan. 17. The Institute is scheduled.

Merits of the Proposal

The merits of the proposal are obvious. For future Institutes the information available would show how the brochure was prepared this year, and subsequent planning would be much simplified.

This information would also be helpful to the people who must send out the press releases on the Institute, and to those who must prepare the general calendar. Should a major in Journalism in the future want to repeat this procedure in a special-problems course, he would have the information his advisor would need. Perhaps in the future too, the design of brochures would be assigned to a journalism class and thus relieve the committee of the task.

It is my hope too that better planning would prompt the committee to act more efficiently in organizing the Institute on National Affairs. A complete schedule would be available earlier so that publicity could be handled more effectively.

Respectfully submitted,

(Signature)

Lise Hurst

Documentation, Mechanics, Diction

18

Documentation

Many scientific papers and reports are derived entirely from firsthand experience (or experiment) and observation. Their authors have few problems arising from the use of printed sources.

Frequently, however, writers may need to supplement their firsthand information by reference to the work of others. Unless the subject is entirely new—and it rarely is—they will need to summarize the literature concerning it. If they need any information derived from other than firsthand observation, they are obliged to indicate its source.

They must, then, be familiar with the techniques and the conventions of *documentation;* that is, "3. The provision of footnotes, appendices, or addenda, referring to or containing documentary evidence in verification of facts or in support of theory in a piece of writing. . . ."—By permission. From *Webster's Third New International Dictionary,* © 1976 by G. & C. Merriam Co., Publishers of the Merriam-Webster Dictionaries.

IMPORTANCE OF MAKING ACKNOWLEDGMENTS

The importance of acknowledging sources is emphasized in the following passage from the *Student's Manual for English 104-105* (1974) prepared by the Department of English at Iowa State University, Section IV:

> In your compositions, whenever you use the phrasing, ideas, or information of some other writer, you must acknowledge your source. Summarizing material from another writer, even though in your own words, does not free you from acknowledging the source. . . .
>
> Unacknowledged use of the information, ideas, or phrasing of other writers is an offense comparable with theft and fraud, and it is so recognized by the copyright and patent laws. Literary offenses of this kind are known as plagiarism . . . .

WHAT TO ACKNOWLEDGE

If you quote exactly, even if only a few words, put the passage within quotation marks (or distinguish it by use of wider margins and often smaller type). Then, especially in scientific writing and reports, indicate the source exactly enough that a reader could find it in a good library.

An exception might be in quoting extremely familiar passages, for example, from Shakespeare or the Bible. If you quote the line "The quality of mercy is not strain'd," you might indicate that it is quoted, but probably not that it is from Shakespeare's *The Merchant of Venice,* Act IV, Scene 1, line 184. You might even omit the quotation marks on the assumption that every literate reader is familiar with the line and doesn't need to be told that it is quoted.

Another exception would be in the writing of an abstract, whose author may use the exact words of the original without quotation marks.

THE NEED FOR ACCURACY

Quote exactly, and completely enough that the original is not misrepresented. "Quoting out of context" is one of the favorite devices of the propagandist and the politician.

If you omit words from a quotation, indicate the omission by inserting three periods if part of a sentence is left out, four periods if one or more sentences are omitted. For example, the quotation of the first part of the excerpt from the *Student's Manual* in the preceding section might look like this:

> In your compositions . . . you must acknowledge your source.

The use of four periods to indicate the omission of whole sentences is illustrated at the end of the first paragraph of the same quotation.

If you insert words in a quoted passage, enclose them in square brackets, thus:

> Unacknowledged use of the information, ideas, or phrasing of other writers [even though not quoted word for word] is an offense comparable with theft and fraud, and it is so recognized by the copyright and patent laws.

If the quotation contains a grammatical mistake, a misspelling, or some other error, reproduce the error exactly, even the most obvious of misprints. Then add the Latin word *sic* in square brackets to inform the reader that the error was in the original source.

Be especially careful not to misrepresent by only partially quoting the original. For example, the quotation below from the second paragraph of the passage from the *Student's Manual* is exact, but the meaning is changed:

> . . . use of the information, ideas, or phrasing of other writers is an offense comparable with theft and fraud. . . .

The statement is true only of *unacknowledged* use.

WAYS OF MAKING ACKNOWLEDGMENTS

Six ways of making acknowledgments are in common use: in the introduction, in the text, in notes, in a list of references, in a modified parenthetical style, and in a bibliography.

In the Introduction

In scientific papers, reports, and theses it is common practice to include in the introduction or preface an acknowledgment of sources used, especially if a few sources have been used extensively. Such an acknowledgment may cover not only printed sources but also facts gathered from interviews or observations of the work of other. The statement on page 276 from the Preface of the first edition of *Medical Writing,* by Morris Fishbein and Jewel F. Whelan, illustrates this practice.

The help of the following employees of the American Medical Association, whose suggestions are also incorporated in this book, is hereby acknowledged with the gratitude of the authors: Marjorie Hutchins Moore, Librarian; Dr. Morton S. Biskind; Mr. F. K. Bryant, and Dr. Paul Nicholas Leech, Secretary of the Council on Pharmacy and Chemistry. . . .

In the Text

Acknowledgment may be made in the text, either just before or just after a quotation. Usually such an acknowledgment will refer only to author, or to author and title. Sometimes it may include complete bibliographical data, as in the following examples:

1. "Any report based on the writing of others should acknowledge the sources used. Not only is it common courtesy (or honesty) to give credit where credit is due, but it is a sign of scrupulousness to tell the source of a statement, so that a reader can judge for himself the evidence it is based on. . . ." —Porter G. Perrin, *Writer's Guide and Index to English,* third edition, Chicago: Scott, Foresman and Company, 1959, p. 374.

2. "Every unoriginal statement must be documented. That means two footnotes must be written if a single sentence contains information from two distinct sources. If a paragraph contains information from a dozen sources, a dozen footnotes appear at the bottom of the page. If several pages are based on one source, just one note is needed, at the end of the discussion." —Gordon H. Mills and John A. Walter, *Technical Writing,* third edition, 1970, Holt, Rinehart and Winston, Inc., New York, Chicago, San Francisco, Atlanta, Dallas, Montreal, Toronto, London, Sydney, page 448.

In Footnotes

Perhaps the commonest practice in formal scientific papers is to indicate sources in footnotes, references to which are made by small "exponent" figures placed slightly above the line at the end of the quotation or passage in which facts derived from sources are used, and after any punctuation.

Just what data should be included in the footnote will depend on circumstances. In their fullest form they will include enough details that readers could find the source in a good library, as they could if they were tracing the following references:

[1]Theodore M. Bernstein, "Windyfoggery," from *The Careful Writer,* pages 480-482, Atheneum, New York, 1965.

[2]J. J. Mahoney, Jr., "New Multiplex for Long Distance Service," *Bell Laboratories Record,* Vol. 42, No. 1, January 1964, pp. 32-38.

The exact order of the data in the footnote is not fixed, except that the author's name is first, the title second. The Christian name or the initials of the author always precede the surname, and do not follow it as in bibliographies. Otherwise, different practices in arrangement, punctuation, use of abbreviations, etc., prevail in different fields. For illustrations of various practices see Section 6.1 under Punctuation, page 304.

If the paper has an appended bibliography of sources as well as footnotes, information about publisher, date, volume, number, etc., need be given only in the bibliography. The footnote reference can be limited to author, or author and title, and page number.

When a paper is sprinkled with frequent footnote references, many devices to secure economy are in common use, such as *ibid.*, "in the same work," and *op. cit.*, "work (already) cited." For such refinements of usage in documentation, consult any of the standard reference works listed on pages 358-360.

Note that in typed and printed matter the footnote at the bottom of the page is usually separated from the text by a line an inch or so long extending from the left margin. When manuscript is being prepared for a printer, it is customary to insert any footnote between two solid lines just after the reference number in the text.

In a List of References

A common practice in technical journals and reports is to print all footnotes at the end in a numbered list in the order in which the citations occur in the text. In the body of the paper the references are made by numbers in parentheses. Here is a typical example from the *Biophysical Journal*:

> *Passage from the text*: . . . It is an indication that the laws of surface tension or membrane elastic tension (6, 7) apply to the cell in these circumstances and at least the tongue of the cell does behave like a fluid-filled membrane. . . .
>
> *Items from the "List of References"*:
>
> 6. Novoshilov, V. V., 1959, The Theory of Thin Shells, translated by P. G. Lowe, P. Noordhoff, Ltd., Groningen, Netherlands.
> 7. Timoshenko, S., 1940, Theory of Plates and Shells, New York, McGraw-Hill Book Company.

Though this method avoids cluttering up the pages with footnotes that few readers will use, it has disadvantages. First, if readers want to know the source, they must waste time consulting a list at the end of the chapter instead of glancing at the bottom of the page. Second, page numbers are omitted—necessarily, since several references to different pages of the same source may occur in the text.

To avoid this second weakness, writers often follow each parenthetical reference in the text with the page number. The foregoing citations would have been more complete if they had been written (6, p. 234; 7, p. 16). The first disadvantage, of course, still remains.

In a Modified Parenthetical Style

A growing number of disciplines use a modified parenthetical style for references within the text. The author's name, the year of publication, and the page number are enclosed within parentheses—like this (Smith, 1969: 455). Notice the placement of the parenthetical reference inside the sentence's terminal punctuation. A reader who wants to check this reference then turns to the bibliography to find the item by Smith dated 1969. If there is more than one item published by Smith in 1969 and listed in the bibliography, the references in the text must be thus: 1969a, 1969b, etc. This method has the advantages of giving the reader more than just a reference number without the reader's having to look beyond the page at hand, and of removing the necessity to include both a list of references and a bibliography.

In a Bibliography

A bibliography, usually at the end of a scientific paper or report, though it may sometimes contain the only acknowledgments of sources, is more likely to be used to supplement footnotes. It is a list, alphabetized according to the first letters of the authors' surnames (see examples on page 252), not arranged in the order of references in the preceding text. It is not an adequate substitute for footnotes or a list of references. The bibliography is appropriate for a list of general references on the subject, or of authoritative books suggested for additional reading, or of all the sources specifically cited in footnotes. Typical entries are included on page 280 at the end of this chapter.

LAYOUT OF QUOTATIONS

If a quotation is less than two lines in length, it is usually inserted in the text and set off by quotation marks. For example, a quotation from *A Manual of Style* of the University of Chicago Press (Eleventh edition, 1949, p. 193), "Quotations of more than two lines in length should be identified by indention. . . ." is so inserted here.

Quotation of more than two lines in length—some authorities say four or five lines—should be indicated by indenting; for example,

> *Quote* (transitive verb) is used most accurately in the sense of exact citation. When less specific reference to something is intended, *cite* is more appropriate. *Quote* (noun), as a substitute for quotation, is considered unacceptable in writing to 85 percent of the Usage Panel.—© 1979 by Houghton Mifflin Company. Reprinted by permission from *The American Heritage Dictionary of the English Language*.

Quotation marks are omitted when a passage is indented.

EXERCISES

A. *Consult at least three sources for the latest information about some recent scientific or technological development or some current political or economic problem. In a paper presenting your findings, quote verbatim from at least two of your sources; paraphrase the other. Include both footnotes and a bibliography.*

Note the use of footnotes and bibliography in the following paper written in 1964.

DISCOVERY OF PETROLEUM IN THE NETHERLANDS AND THE NORTH SEA

Some of the most important discoveries of the century have recently been made in the Netherlands and the North Sea near by. Late in 1959 in Groningen Province in northern Netherlands gas was discovered. The drillers were searching for oil, and had drilled 34 dry wells before the gas strike was made. It was another year before it was realized that a major deposit had been found.

One reliable reference work reports that

> the deposits amounted to at least 150,000,000,000 cu. m. and possibly to as much as 400,000,000,000 cu. m. Annual production was expected to be as much as 8,000,000,000 cu. m., representing more than one quarter of the annual consumption of energy in The Netherlands.[1]

Another report increased these estimates to 523 billion cubic yards (about 680 billion cubic meters) and reported talk of 785 billion cubic yards (about 925 billion cubic meters).[2]

One major difference between the development of this new field in Europe and the development of any similar field in the United States is that the Dutch government, and not private interests, is in control. Though the big gas discovery was made by a company owned jointly by the Dutch government, Standard Oil (New Jersey), and the Royal Dutch Shell Oil Company, and called the Netherlands Oil Company, "The Government assures itself of close to 70 percent of the new company's profits through its direct and indirect share of the enterprise."[3] And, contrary to practice in the United States, the land owner receives "only a nominal rental—two and a half times what the two acres presently in use would ordinarily yield from crops. It comes to about $250 a year."[4]

The effects of this discovery upon the economy of the Netherlands are difficult to estimate.

> For example, the amount of gas now used for heating Dutch homes is negligible, but by 1975 it is expected to rise to 75 percent of the househeating market.[5]

1. *Encyclopedia Britannica,* 1963 Book of the Year, "Netherlands," page 596.
2. *U.S. News & World Report.* "Big 'Find" in a Small Country—the Dutch Strike It Rich," May 20, 1963, page 79.
3. *Ibid.,* page 81.
4. James H. Winchester, "Holland Strikes It Rich in Natural Gas," *Reader's Digest,* August, 1964, page 84.
5. *Ibid.,* page 86.

Eventually pipe lines will distribute gas from the Groningen field to much of Western Europe.

In the North Sea, oil has been discovered, and already extensive geophysical explorations are being carried out, not only off the Netherlands coast but off the east coast of England and the west coasts of Germany, Denmark, and Norway as well. "A major North Sea exploration play is no longer merely a prediction of geologists. It's here."[6] Several companies are at work, among them Mobil Oil AG, Gulf Oil, American International, Atlantic Refining Company, and Global Marine Exploration Company, which have reached exploration agreements with countries bordering the North Sea.

> Under its waters, geologists believe, lie gas pockets whose reserves may exceed even those of the mammoth gas fields under the Texas Panhandle, hitherto the largest known.[7]

It is obvious that "tiny Holland is destined to play a giant's role in Western Europe's new energy age."[8]

BIBLIOGRAPHY

1. *Encyclopaedia Britannica,* 1963 Book of the Year, "Netherlands"
2. *Oil and Gas Journal,* "Two offshore drilling programs in works," May 6, 1963, pages 94, 96
3. *U. S. News & World Report,* "Big Find in a Small Country—the Dutch Strike It Rich," May 20, 1963, pages 79-81
4. Winchester, James H., "Holland Strikes It Rich in Natural Gas," *Reader's Digest,* August, 1964, pages 83-86

6. *Oil and Gas Journal,* "Two offshore drilling programs in works," May 6, 1963, page 94.
7. James H. Winchester, *op. cit.,* page 85.
8. *Ibid.,* page 86..

B. *Observe the practices followed in acknowledging sources in scientific papers in your own field, and write a report on what you find.*

C. *Update the report printed under Exercise A, above using 1980 sources and changing parts of it where necessary.*

19

Mechanical Aspects of Style

THE MATERIAL OF THIS CHAPTER

In this chapter the term "style" refers to those mechanical aspects of the written or printed language that conform to rules and established usage. The word is not used in its literary sense of a manner of expression characteristic of an individual, a period, or a school. The chapter is limited to analysis of usage in six areas: abbreviations; hyphenation of compound adjectives; writing of numbers; capitalization; methods of indicating titles of books, journals, reports, etc.; and punctuation.

If all scientific and technical matter were printed, perhaps such mechanical aspects of style could safely be left to the editor and the printer. It is part of their business to edit copy submitted for publication. But most scientific and technical writing is probably in typewritten form and will not be scrutinized by an editor. Without a knowledge of those rules that apply to typed as well as to printed text, you may turn out faulty manuscript.

Many of these rules are so consistently observed that they need not be listed here. You know that sentences begin with a capital letter and end with a period; that divisions of a word at the end of a line are indicated by a hyphen; that *it's* cannot be substituted for *its*.

Unfortunately, however, many usages are not settled. Some of the variants occur in all kinds of writing; others, most frequently in only one kind. In this chapter—except in the section on Punctuation—are listed only those rules which apply frequently in scientific writing and about which authorities do not agree.

In many of these areas it is impossible to give a single rule that states the only right practice or even the best practice. Usage is often divided; authorities often do not agree. What is established practice in one field may be unacceptable in another field. Remember, then, that these usages do vary. Observe the prevailing usage in your own field, and then practice that usage in your own writing.

BIBLIOGRAPHY

Many reference works, dictionaries, textbooks, handbooks, and style books constitute the "authorities" (the word is in quotation marks because many of them are not always dependable) to which you can refer for guidance. Some of them summarize practice in one publishing house or in one field; some occasionally seem to indicate merely the author's personal preferences. Too rarely do they present evidence that the generalizations are based on studies of actual usage.

If a style book is available which codifies usage in your own field, you should of course be familiar with its rules and obey them when you write. A doctor, for example, should refer to the style book of the American Medical Association. A biologist should refer to the C.B.E. (Council of Biological Editors) Style Book. If no such guide is available in your field, or if the one available does not rule on the point in question, you should have at least one good general reference at hand, and preferably two or three, so that you can find out whether the authorities disagree. No one authority can be recommended as a safe guide on all matters of style, especially in the scientific field. A selected list of those that can be recommended (with reservations) is printed in the bibliography at the end of this book.

ABBREVIATIONS

The forms and uses of abbreviations vary so much in different fields that it is difficult to generalize about them.

Reasons for Their Use

Abbreviations are used for several reasons:

1. Some terms appear only as abbreviations; for example, *Mr.*, *Jr.*, *a.m.*, B.C.
2. Some abbreviations (especially acronyms, formed from initial letters of words) are more common and more easily understood than the full forms; for example, DDT, polio, NATO.
3. Using abbreviations saves a writer's time. It takes less time to write *emf* than "electromotive force," *cc* than "cubic centimeter."
4. Abbreviations take less space than the fully spelled-out words.
5. Writers use abbreviations because they have the habit.

In technical and scientific writing any of these reasons except the last may justify the use of an abbreviation. But the reasons for *not* using them may be stronger:

1. Readers may not be familiar with the abbreviation. If not, communication stops while they try to translate the unfamiliar symbol.
2. The saving of the writer's time may be more than offset by the wasting of the reader's time. If the writer saves 5 seconds by writing *mep* for "mean effective pressure" and the reader wastes 5 minutes figuring out the meaning of the abbreviation, there is a net waste of 4 minutes and 55 seconds.
3. The saving in space may be too slight to justify the time saved by the writer. Even when abbreviations are used freely, the saving in space will rarely exceed 2 percent of the whole paper, and the saving in time even for the writer familiar with them will probably be less.

Especially in reports intended for readers whose knowledge of technical details is incomplete, you should question whether abbreviations do not impede communication. If you decide to use them, perhaps you should include a list, with meanings spelled out.

It is good practice to write out in full at first a term you intend to abbreviate, and follow with the abbreviation in parenthesis. For example "The standard operating procedure (SOP) has been long established." The weakness of this practice is that the reader may not find or remember your first use of the abbreviation.

Extent of Their Use

The overuse of acronyms has become a mark of modern style in some fields, especially government writing. The second edition of the *Acronyms and Initialisms Dictionary,* by Frederick G. Ruffner, Jr., has 45,000 entries. *Newsweek,* on March 11, 1966, commented on "FASGROLIA: the Fast Growing Language of Initialisms and Acronyms," calling it "an insider's shorthand, a technocrat's jargon, a vast and sprawling collection of abbreviations, sometimes useful, sometimes witty, often confusing and grotesque, which is invading the nation's speech." Most of us have encountered some of its terms in our reading—QSO, RA, rbi, SHAPE, UAR—and have been frustrated by not knowing their meanings. Many of them, if widely enough used, are listed in good recent dictionaries.

On the other hand, in some exposition almost no abbreviations are used except of terms that rarely appear fully spelled out. Elsewhere, especially in technical writing, abbreviations occur frequently. In nonscientific, nontechnical matter, any except the most common abbreviations are usually avoided.

Even in strictly technical matter, the extent of the use of abbreviations varies greatly. In some fields, however, the trend is toward fewer abbreviations. For example, in business letters many firms no longer abbreviate even the names of states and months, or of "street" and "avenue." Since the excuse for writing "St." and "Oct." is to save time and space, and either may be interpreted as slightly discourteous, good practice requires a minimum of abbreviation.

Forms of Abbreviations

The spelling and punctuation of abbreviations are not standardized. Lists published only a few years ago employed many capital letters and almost always included periods. Such abbreviations as B.T.U. and E.M.F. (or B.t.u. and e.m.f.) were widely used and are still recommended by many "authorities." The newer practice, adopted by the American Standards Society and the leading engineering and scientific societies, uses capitals sparingly and omits periods except where the omission might be confusing—as after *in.* (inch) and *bar.* (barometer).

The confusing diversity of forms of abbreviations for terms in common use is illustrated by the following list:

British thermal unit B.T.U.: — B.t.n. — b.t.u. — Btu. — B — Btu
Electromotive force: E.M.F. — e.m.f. — EMF — emf
Diameter: diam. — dia. — di. — d. — D. — diam
Kilogram: K — kilo. — kg. — kgm. — kilo — kg

Plurals of Abbreviations

The present recommended practice is to use the same form of abbreviations for both singular and plural. The following examples are typical:

3 m – 4 kg – 2 in. – 5 yd – 2 hr – 3 gal – 5 bu – 7 lb

In literary references, footnotes, and bibliographical references and occasionally elsewhere it is common practice, however, to add *s* to form the plural of terms such as the following:

vols. I and II	figs. 2 and 3
chaps. ii and iii	nos. 5 and 6
secs. 4 and 5	arts. 1 and 2

A few plural abbreviations are formed irregularly, like *ll* for *lines* and *pp* for *pages,* usually without periods.

A.S.A. Abbreviations for Scientific and Engineering Terms

In 1941 the American Standards Association designated the following list of abbreviations as standard and it is now approved by the leading scientific and engineering societies. The list is published by the American Society of Mechanical Engineers. The introductory statement says that "these forms are recommended for readers whose familiarity with the terms used makes possible a maximum of abbreviations. For other classes of readers editors may wish to use less contracted combinations made up from this list. For example, the list gives the abbreviation of the term 'feet per second' as 'fps.' To some readers 'ft per sec' will be more easily understood."

absolute .... abs
acre .... spell out
acre-foot .... acre-ft
air horsepower .... air hp
alternating current as adjective .... a-c
ampere .... amp
ampere-hour .... amp-hr
amplitude, an elliptic function .... am.
Angstrom unit .... A
antilogarithm .... antilog
atmosphere .... atm
atomic weight .... at. wt
average .... avg
avoirdupois .... avdp
azimuth .... az or *a*

barometer .... bar.
barrel .... bbl
Baume .... Be
board feet (feet board measure) .... fbm
boiler pressure .... spell out
boiling point .... bp
brake horsepower .... bhp
brake horsepower-hour .... bhp-hr
Brinell hardness number .... Bhn
British thermal unit .... Btu or B
bushel .... bu

calorie .... cal
candle c
candle-hour .... c-hr
candlepower .... cp
cent .... c or ¢
center to center .... c to c
centigram .... cg
centiliter cl
centimeter .... cm
centimeter-gram-second (system) .... cgs
chemical .... chem
chemically pure .... cp
circular .... cir
circular mils .... cir mils

coefficient .... coef
cologarithm .... colog
concentrate .... conc
conductivity .... cond
constant .... const
continental horsepower .... cont hp
cord .... cd
cosecant .... csc
cosine .... cos
cosine of the amplitude, an elliptic function .... cn
cost, insurance, and freight .... cif
cotangent .... cot
coulomb .... spell out
counter electromotive force .... cemf
cubic .... cu
cubic centimeter .... cc, cu cm. cm^3 (liquid, meaning milliliter, ml)
cubic foot .... cu ft
cubic feet per minute .... cfm
cubic feet per second .... cfs
cubic inch .... cu in.
cubic meter .... cu m or m^3
cubic micron .... cu μ or cu mu or μ^3
cubic millimeter .... cu mm or mm^3
cubic yard .... cu yd
current density .... spell out
cycles per second .... spell out or c
cylinder .... cyl

day .... spell out
decibel .... db
degree .... deg or °
degree centigrade C
degree Fahrenheit .... F
degree Kelvin .... K
degree Reaumur .... R
delta amplitude, an elliptic function .... dn
diameter .... diam
direct-current (as adjective) .... d-c
dollar .... $
dozen .... doz
dram .... dr

efficiency .... eff
electric .... elec
electromotive force .... emf
elevation .... el
equation .... eq
external .... ext

farad .... spell out or f
feet board measure (board feet) .... fbm
feet per minute .... fpm
feet per second .... fps
fluid .... fl
foot .... ft
foot-candle .... ft-c
foot-Lambert .... ft-L
foot-pound .... ft-lb
foot-pound-second .... fps
foot-second (see cubic feet per second) ....
franc .... fr
free aboard ship .... spell out
free alongside ship .... spell out
free on board .... fob
freezing point .... fp
frequency .... spell out
fusion point .... fnp

gallon .... gal
gallons per minute .... gpm
gallons per second .... gps
grain .... spell out
gram .... g
gram calorie .... g-cal
greatest common divisor .... gcd

haversine .... hav
hectare .... ha
henry .... h
high-pressure (adjective) .... h-p
hogshead .... hhd
horsepower .... hp
horsepower-hour .... hp-hr
hour .... hr
hour (in astronomical tables) .... h
hundred C
hundredweight (112 lb) .... cwt
hyperbolic cosine .... cosh
hyperbolic sine .... sinh
hyperbolic tangent .... tanh

inch .... in.
inch-pound .... in-lb.
inches per second .... ips
indicated horsepower .... ihp
indicated horsepower-hour .... ihp-hr
inside diameter .... ID
intermediate-pressure (adjective) .... i-p
internal .... int

joule .... j

kilocalorie .... kcal
kilocycles per second .... kc
kilogram .... kg
kilogram-calorie .... kg-cal
kilogram-meter .... kg-m
kilograms per cubic meter .... kg per cu m or kg/m³
kilograms per second .... kgps
kiloliter .... kl
kilometer .... km
kilovolt .... kv
kilometers per second .... kmps
kilovolt-ampere .... kva
kilowatt .... kw
kilowatt-hour .... kwhr

Lambert L
latitude .... lat or ϕ
least common multiple .... lcm
linear foot .... lin ft
liquid .... liq
lira .... spell out
liter l
logarithm (common) .... log
logarithm (natural) .... $\log_e$ or ln
longitude .... long. or y
low-pressure (as adjective) .... l-p
lumen l
lumen-hour .... l-hr
lumens per watt .... lpw

mass .... spell out
mathematics (ical) .... math
maximum .... max
mean effective pressure .... mep
mean horizontal candle-power .... mhcp
megacycle .... spell out
megohm .... spell out
melting point .... mp
meter .... m
meter-kilogram .... m-kg
mho .... spell out
microampere .... μa or mu a
microfarad .... μf
microinch .... μin.
micromicrofarad .... $\mu\mu$f
micromicron .... $\mu\mu$ or mu mu
micron .... μ or mu
microvolt .... μv
microwatt .... μw or mu w
mile .... spell out
miles per hour .... mph
miles per hour per second .... mphps
milliampere .... ma
milligram .... mg
millihenry .... mh
millilambert .... mL
milliliter .... ml
millimeter .... mm
millimicron .... mμ or m mu
million .... spell out
million gallons per day .... mgd
millivolt .... mv
minimum .... min
minute .... min
minute (angular measure) ....
minute (time) (in astronomical tables) .... m
mole .... spell out
molecular weight .... mol. wt
month .... spell out

National Electrical Code .... NEC

ohm .... spell out or Ω
ohm-centimeter .... ohm-cm
ounce .... oz
ounce-foot .... oz-ft
ounce-inch .... oz-in
outside diameter .... OD

parts per million .... ppm
peck .... pk
penny (pence) .... d
pennyweight .... dwt
peso .... spell out
pint .... pt
potential .... spell out
potential difference .... spell out
pound .... lb
pound-foot .... lb-ft
pound-inch .... lb-in.
pounds per brake horse-power-hour .... lb per bhp-hr
pounds per cubic foot .... lb per cu ft
pounds per square foot .... psf
pounds per square inch .... psi
pounds per square inch absolute .... psia
power factor .... spell out or pf

quart .... qt

radian .... spell out
reactive kilovolt-ampere .... kvar
reactive volt-ampere .... var
revolutions per minute .... rpm
revolutions per second .... rps
rod .... spell out
root mean square .... rms

secant .... sec
second .... sec
second (angular) .... ″
second-foot (see cubic feet per second)
second (time) (in astronomical tables) .... s
shaft horsepower .... shp
shilling .... s
sine .... sin
sine of the amplitude, an elliptic function .... sn
specific gravity .... sp gr
specific heat .... sp ht
spherical candle power .... scp
square .... sq
square centimeter .... sq cm or cm^2
square foot .... sq ft
square inch .... sq in.
square kilometer .... sq km or km^2
square meter .... sq m or m^2
square micron .... sq μ or sq mu or μ^2
square millimeter .... sq mm or mm^2
square root of mean square .... rms
standard .... std
stere .... s

tangent .... tan
temperature .... temp
tensile strength .... ts
thousand .... M
thousand foot-pounds .... kip-ft
thousand pound .... kip
ton .... spell out
ton-mile .... spell out

versed sine .... vers
volt v
volt-ampere .... va
volt-coulomb .... spell out

watt .... w
watt-hour .... whr
watts per candle .... wpc
week .... spell out
weight .... wt

yard .... yd
year .... yr

Exercises

A. *Observe the usage in writing abbreviations in a reputable journal in your field, and report your findings.* Include answers to the following questions: (a) To what extent and how consistently are abbreviations used? (b) Are the forms used the ones recommended by the American Standards Association? (c) Are the same abbreviations used for both singular and plural? (d) What terms are regularly spelled out in full that might have been abbreviated? (e) Did any of them slow you down in your reading? (f) How much space is saved by the use of abbreviations?

B. *Report on the extent of the use of acronyms in a newspaper or a weekly news magazine, such as Newsweek, Time, U. S. News and World Report. Were any used without explanation?*

HYPHENATION OF COMPOUND ADJECTIVES

Two or more words that combine to form a unit modifier preceding the word or words modified should be hyphenated; for example,

> 10,000-bottle capacity, lower-income group, 500-lb-per-day chlorinator, aeronautical-research facilities

Nearly all the authorities give this rule, though some of them admit exceptions. It is observed with considerable uniformity by editors of magazines and newspapers and by authors of books in general fields, as well as by editors of technical and scientific publications.

The hyphenation of compound adjectives, as Summey noted in *Modern Punctuation,* has "the virtue of enabling the writer to turn an ordinary phrase into an adjective, as in 'large-scale production' and 'limited-liability companies.'" "The value of the hyphen is that it enables the reader to see instantly that the words so joined are a unit modifier and not separate modifiers. Even though no real ambiguity would result from the omission of the hyphen, its use always makes for quicker and easier comprehension. Sometimes the omission of the hyphen will result in completely different meaning, as in these examples:

a deep-green sea	a deep green sea
five-dollar bills	five dollar bills
12-hour intervals	12 hour intervals
dry-vacuum tank	dry vacuum tank
stock-solution tank	stock solution tank

The compounding of adjectives has become extremely frequent in scientific and technical writing. Some critics have objected strenuously to the growth of the practice, notably Lord Dunsany in an article on "Decay in the Language," with its example "England Eleven Captain Selection Difficulty Rumor." Absurd as the example is, it would have been rendered intelligible by the insertion of four hyphens. T. A. Rickard, in *Technical Writing,* says that the hyphen is "a means of condensing the phraseology; and so long as the meaning is not obscured it is a useful device, but the technical writer must be careful not to addict himself to an excessive use of the hyphen because it leads to the acquisition of a style that may become both ugly and obscure. . . . The moral is to avoid compounding, and to indicate the meaning of the unavoidable compounds by placing the hyphens carefully."

The value of the hyphen as an aid to comprehension can be seen by comparing phrases with and without the mark:

forced draft chain or traveling gate stokers	forced-draft chain- or traveling-grate stokers
a piston type pump	a piston-type pump
a public utility plant	a public-utility plant
V engine powered Hurricane	V-engine-powered Hurricane
6-in. diameter hardwood rollers	6-in.-diameter hardwood rollers

Exceptions to the General Rule

Some of the reference books list exceptions to the general rule for hyphenating compound adjectives. Hyphens are generally regarded as unnecessary in the following kinds of expressions:

1. Adjectives formed of two proper nouns having their own fixed meaning; *Old Testament times, New York subways.*
2. Foreign phrases; for example, *laissez faire policy.*
3. Chemical terms; for example, *sodium chloride solution.*
4. Compound adjectives following the noun they modify; for example, *sea of a deep green, an accident little expected, a name well known. The Standard Handbook for Secretaries,* however, rules that such a combination should be hyphenated when it retains its force as a one-thought modifier; for example, an *existence that became hand-to-mouth, arrows that are hand-tipped.*

Other exceptions made by some authorities but not commonly agreed upon are the following:

1. Comparatives or superlatives with the first word ending in *er* or *est;* for example, *a higher priced article, the deepest toned instrument.*
2. Combinations of adjective and adverb or of adverb and participle when the ommission would result in no ambiguity; for example, *an ever increasing flood, a highly developed species.*
3. Ordinary names used as modifiers; for example, *income tax returns, real estate values, night letter rates.*
4. A three-word modifier in which the first word modifies the second, but not the second and third together; for example, *fairly well defined course;* but *nearly right-angle bend.*

The omission of the hyphen between an *ly* adverb and a following adjective or participle (exception no. 2) is approved by many authorities and has much good usage to recommend it. Since the *ly* adverb could not possibly be mistaken for a separate modifier of the following noun, the hyphen is superfluous.

The rule most easily followed and nearly always justifiable in scientific and technical writing is to use the hyphen in all compound adjectives preceding nouns.

Suspension Hyphen

When two or more compound adjectives preceding a noun have a common base word, the base word is usually omitted in the first adjective, although the hyphen (called a "suspension hyphen") remains; for example,

a fourth- or fifth-grade lesson
light- or attack-bomber class
five- or three-cycle breakers
2¼- by 4¼-inch block

The suspension hyphen is so frequent in scientific prose that every writer should be familiar with its use.

Exercise

Insert hyphens in the following phrases where they are needed:

second ranking minority member
a safe deposit box
elevated railroad tracks
New York Washington corridor
last minute offer
South Capitol Street entrance
oil burning electric power plants
balance of payments deficits
San Francisco New York air mail
1/16 inch diameter copper coated steel wire
red and white one and a half inch WIN button
our first in person view
single pole switch
slowly revolving motor
heavy truck driver
all inclusive two week round trip package
never to be forgotten chills
twenty first century edifice
black bird trap
an 8 by 10 by 12 foot space
a space 8 by 10 by 12 feet
four and a half year old daughter
a mechanical engineering student
a beautiful black and white cotton knit suit
high pitched Ed Wynn giggle
power in reserve ease
devil may care man to man fashion
Police Department order
most beautiful residences
dark blue pin striped suit
thirty two year old Elizabeth Holtzman
red and white tulips
telephone book size document
a Chicago textbook publisher
an economics textbook publisher
a well coordinated athlete
green and white striped corded jacket
a blue leather chair
third or fourth rate burglary
4½ inch diameter brass tubing
25 ohm variable coil resistance
printed circuit diagram
potassium bromide solution
seven large capacity hydrogen cooled synchronous condensers
a legitimate national security measure
automatic depth of field indicator
independent front and rear suspensions
four wheel power disc brakes
Bank of England type
a New York State chartered bank
a 12 to 14 hour a day job
investment banking firm

THE WRITING OF NUMBERS

When numbers should be spelled out and when they should be written as figures are unsettled questions. Authorities do not agree on the rules, and publications do not conform to any single practice. Publishers and printers seem to make their own rule and follow the usages they prefer.

The Trend of Usage

Unquestionably the general trend in writing numbers is toward the use of figures, especially in scientific and technical contexts. But no authority recommends the use of figures for all numerical values. Conflicting logical and aesthetic principles, as well as the weight of established custom, have prevented such consistency. Many think that figures spoil the look of the printed page, and approve their use only when custom clearly requires it. Others have become accustomed to the spelled-out forms in certain situations and oppose the use of figures there. Actually, few readers are conscious of anything unusual in a text where all numerical values are in figure form.

Quantitative Expressions and Aggregates

It is helpful to classify all numbers as *quantitative expressions* and *aggregates.* Quantitative expressions are those including some unit of measurement, such as *yard, minute, gram, gallon, volt, and acre.* Aggregates are those expressions resulting from the adding up of items and not including a unit of measurement. The distinction is evident from the following examples:

Quantitative expressions	Aggregates
board 16 feet long	six men
3-year-old child	five votes
40 bushels per acre	nine countries
angle of 10 degrees	one reason
2-gallon can	three stamps
6 acres in the field	four fields

In scientific and technical writing the use of figures in all quantitative expressions (with some exceptions noted below) is so common that it can be recommended as a general rule.

Usage in writing aggregates varies. Common practice is to draw a line somewhere in the scale of numbers and spell out those below the line and use figures for those above (unless some other rule requires that it be a figure.) Usually the line is between 10 and 11. Some authorities recommend spelling out values not requiring more than three or four words to write, like "one hundred and fifty-two." Some recommend figures for small aggregates occurring in a passage with larger numbers.

The distinction between quantitative expressions and aggregates is a convenient one and makes the choice between figures and spelled-out words easy. A single rule (with a few exceptions) can be followed:

Use figures to express all numerical values except small aggregates.

Some Special Cases

A few kinds of numerical expressions deserve special attention.

Approximations

The tendency at present is to use figures for both approximations and exact numbers; for example,

an army of 1,000,000 men
more than $6,000,000
80,000 engineers
approaching 1,000,000,000 kwhr
20 years

Large Numbers

Round numbers above a million are now frequently written as a combination of figures and words; for example,

11 billion dollars
employment of ½ million
1¼-billion-dollar program
6¼ million million million dollars
$5 billion
$1,100 million
7.5 million

A Number Beginning a Sentence

A number at the beginning of a sentence should be spelled out. Such sentences are common and unobjectionable, though some stylists recommend revision to avoid beginning with numbers. Thus the sentence

Fifty men were employed to complete the job on schedule.

can be worded

To complete the job on schedule, 50 men were employed.

Two Numbers Coming Together

The first of two numbers occurring without an intervening word is usually spelled out unless it is a large aggregate; for example,

fifty 6-inch boards

But some authorities recommend using figures for both numbers when the first is more than 100; for example,

120 6-inch boards

Fractions

The commonest practice is to spell out small fractions when they do not appear as part of a mathematical expression or in combination with a unit of measurement; for example,

one-third of the area
about one-tenth their value
with three-fourths load

½-inch pipe
⅝-in.-outside-diameter tubing
¼-inch screenings

Occasional variations from this usage, however, indicate the trend toward the more consistent use of figures; for example,

about ¼ to ½ division
about ¾ mile
⅔ to ¾ of the imbedment

Fractions with larger denominators than 10 are usually written as figures. When the fractions are nouns; they are usually followed by the last two letters of the ordinal; when they modify a noun, they stand alone; for example,

1/80th of a second
1/21st of the cost
1/22nd of the amount

1/16 inch
1/100 horsepower
1/64 acre

MIXED NUMBERS

Mixed numbers are written as figures except at the beginning of a sentence; for example,

Use 1½ cups of flour.
One and one-half cups of flour are used.

DECIMAL VALUES

Decimal values are always written as figures. If the value is less than 1, a zero precedes the decimal point.

0.10, 6.015

SINGULAR OR PLURAL UNIT OF MEASUREMENT

The unit of measurement following a decimal or fraction is singular if the value is 1 or less, plural if it is more than one; see the following examples:

¾ second, *but* 1¾ seconds
0.077 decibel, *but* 4.43 decibels

ORDINALS

Small ordinals are usually spelled out: large ones are in figure form followed by *st, nd, rd, th;* for example,

the eighth patient
the eighteenth century (or 18th)
Tenth Avenue (*but* 42nd Street)

SMALL AGGREGATES IN FIGURE FORM

Exceptions to the usual practice of spelling out small aggregates are the following:

Numbers in mathematical context; for example,

The first integer times 2 equals 8.
a proportion of 2 to 5

Numbers designating pages, paragraphs, lines, verses, etc.; for example,

page 5 verses 6 to 8 figure 3 line 7
paragraph 6 item 6 question 2 No. 4

Exercises

A. *In the following sentences or parts of sentences, decide which values might properly be written as figures.*

1. Working twelve-hour shifts, seven days a week, some workers pull in a thousand dollars a week.
2. When not more than ten units of insulin are required in twenty-four hours, the entire dose may be taken at once, before breakfast, or five before breakfast and five before the evening meal.
3. Only three wells were producing two thousand barrels each; twenty-five were producing one thousand each; and sixty were pumping about a hundred barrels each.
4. He paid for thirty five-cent stamps with two one-dollar bills.
5. He asked us to refer to Figure nine on page ninety-six of Chapter fourteen and to Graph sixteen of Article six, Chapter two.
6. Between the sixth and twenty-first floors the elevator stopped four times.
7. There were three six-room houses, five four-room houses, and three two-room cottages, and they were built by twenty men.
8. The eighth and tenth groups contained three and four items respectively.
9. He may have the fact that $x - 3 = 7$ and wish to find the value of x. He notes that if he adds three to both sides of the equation he will obtain x = ten.
10. We shall still use the fact that two dollars plus two dollars are four dollars, but not that two raindrops added to two raindrops are four raindrops.
11. Ford Granada–two-door, fourteen-eighteen mpg: city/eighteen-twenty-six mpg: highway.
12. More than thirty million dollars of contracts. . . .
13. Two weeks after his death. . . .
14. At four o'clock in the morning more than ten fifty-year-old trolley cars began trundling out of the barn.
15. Trucks help move more than three out of every four tons.
16. A one or two dollar cut per barrel is anticipated.
17. Brown, thirty, nine years ago was sentenced to a term of twenty years in prison.
18. The odds against his winning were about two thousand to one.
19. They are building five tanker berths and forty-four oil storage tanks.
20. The two-act, seventeen-scene opera required eleven changes of costume.

B. *Observe the usage in writing numerical values in a reputable journal in your field and report your findings, with examples and conclusions.*

CAPITALIZATION

A few of the many rules for use of capital letters apply especially to the composition of scientific and technical material.

Titles, Heads, Subheads

In titles, heads, and subheads the general practice is to use either all capitals or initial capitals for the first word and for all other words except articles, prepositions, and conjunctions. Some authorities recommend an initial capital letter for any word of four or more letters and for shorter words if they are stressed; for example,

Decide Before not After Doing
Standards In Business and Out

Legends or Captions

In the writing of legends or captions accompanying figures, drawings, illustrations, etc. usage is divided. You should follow the practice of the leading journals in your field. Some publications set such legends entirely in small capitals, some in small capitals with large capitals for initial letters of important words, some with an initial capital only for the first letter of the first word. For typed manuscript either of the following forms is acceptable:

Figure 1. Bar chart drawn on a horizontal plane
Figure 1. Bar Chart Drawn on a Horizontal Plane

Outlines and Tabulations

The prevailing practice in scientific and engineering matter is to capitalize only the first word of each item of an outline or tabulation. If the tabulation is merely a list of words or phrases, lowercase letters may be used throughout.

Tables of Contents

In the items of tables of contents it is customary to use initial capitals for all important words, in both main heads and subheads. If the tables are very detailed, minor subheads may have initial capitals for only the first word of each item. Occasionally main headings will be typed entirely in capital letters.

Common Nouns Used with Date, Number, or Letter

Common nouns followed by a date, number, or letter are sometimes capitalized, sometimes not. Such expressions as *act of 1928, article II, appendix C, figure* 7, and *collection* 6 may be written with initial capitals (*Act of 1928,* etc.) or without. The trend seems to be toward the use of lowercase initial letters in such items.

Derivatives of Proper Nouns

Words derived from proper nouns tend to drop the initial capitals when they have developed a specialized meaning not suggesting their origin. Such terms as *india ink, castile soap, diesel engine,* and *bunsen burner* are now commonly written without capitals. Even such terms as *americanize, mendelian, freudian,* and *jacksonian* are not capitalized in American Medical Association usage.

Capitalization After a Colon

Usage is divided on capitalization after a colon, but the trend seems to be toward the use of lowercase letters. *A Manual of Style* (University of Chicago) rules that "the first word after a colon is not to be capitalized when it introduces an element that is explanatory or logically dependent upon the preceding clause."

If the colon precedes a series of items set in tabular form and numbered or lettered, however, each item will usually be capitalized; for example,

In criticizing a definition ask those questions:

a. Does the supposed definition include term, genus, and differentiate?
b. Is the genus small enough?

If the series consists merely of a list of words or other items set in a column but not numbered or lettered, the items usually will not be capitalized; for example,

The following terms should always be written in full when they occur alone:

acre
boiler pressure
coulomb
day, etc.

When such items consist of several words, the first word may be capitalized; for example,

The following information should be included, in this order:

Name of the author
Title of publication
Number of the bulletin
Name of the department
Data

Points of the Compass

Words designating points of the compass are capitalized only when they indicate geographical areas; for example,

the South, the Middle West, the Far East, North Africa, south of Chicago, western San Francisco, eastern Persia, northern China

Names of Public Places

Names of public places are capitalized; for example,

Canal Zone, Flushing Meadow, Bedloe's Island, Pennsylvania Station

Exercises

A. *Report on the capitalization in a reputable journal or textbook in your field. Make what generalizations you can about the usages you find, especially those indicated by the headings in the section on capitalization. Include illustrative examples.*

B. *Decide which words should be capitalized in the following headings:*

what is a law?
what may be taken
student ratings of faculty
determining per capita costs
percentage of urban population
should a pupil add up or down a column?
decide before not after doing
out-of-town exchanges
postal rates on 16-mm films
world in all-out war

C. *Decide which words should be capitalized in each of the following phrases:*

assistant to president carter
new national gallery of art
the veteran boxes association
the metropolitan opera house
sergeant major hascarni
the english channel
the french coast
the fourth precinct
bureau of internal revenue offices
gold medal flour
battle of the bulge
sermon on the mount
the milky way
the big dipper
at the south pole
"I could have danced all night"
the senate judiciary committee
the american flag
the joint chiefs of staff
the distinguished flying cross
army, navy, or air force
fifth congressional district
a pullman car
a ford or a chevrolet sedan
forty-second street
du pont house paint
the city of detroit
in chapters I and IV
at camp pendleton on the ocean front
attended a military academy
the empire state building
in any state capital
a wall street banking firm
the white house aides
supreme court justice douglas
laurel avenue no 511

METHODS OF INDICATING TITLES OF BOOKS, MAGAZINES, ETC.

Four methods of indicating the titles of books, magazines, and other units of composition have both authority and good usage to uphold them:

1. Printing in italics (underlining in typed or handwritten manuscript)
2. Inclosing in quotation marks
3. Capitalizing initial letters of principal words
4. Using capitals and small capitals (all capitals in typed manuscript)

Some of the authorities say that one method only is to be used; some specify one method for one kind of title, another method for another kind; some say that the writer is free to choose the method preferred.

Italics

Titles of books, magazines, reports, pamphlets (or any unit separately printed and bound) when used in text matter are commonly italicized, though some publications reserve italics to distinguish titles of magazines and use quotation marks for other titles.

Quotation Marks

Though used by some publications to distinguish all titles, quotation marks more commonly distinguish chapter titles and headings in books; titles of articles, editorials, etc. in magazines; titles of short poems and musical compositions; in other words, titles of any unit not separately printed and bound. The use of italics to distinguish the title of a printed book or magazine and of quotation marks to distinguish titles of a part of the book or magazine is desirable.

Initial Capitals Only

The Government Printing Office *Style Manual* requires merely initial capitals of principal words of titles of all kinds, either in text or in bibliographical lists or footnotes. Some newspapers also observe this rule. In bibliographical lists, footnotes, tables, or other matter where the use of italics or quotation marks might be awkward or inconvenient, distinguishing titles by initial capitals only is common.

Capitals and Small Capitals

A special practice of many publications is to set their own title, when it occurs in text, in capitals and small capitals. In typed letters the practice of using all capitals is also common, especially in the letters of publishers mentioning titles of their own books.

Summary

Though conservative usage prefers either *italics* or quotation marks for titles, weighty authorities and considerable reputable usage favor treating titles like other proper nouns and capitalizing merely the initial letters of the principal words.

PUNCTUATION

The marks of punctuation are a comparatively recent innovation. Not until after the invention of the printing press were they used generally, and not until the eighteenth century were they used with any consistency. Before that, just as writers spelled as they pleased, so they punctuated as they pleased. Nobody had attempted to formulate rules.

Usage in punctuation is still unsettled. Few rules concerning it are universally observed. As will be evident in the following sections, two or more "right" ways of punctuating a sentence are often established in usage. About the only safe generalization possible is that the general trend is toward less punctuation, especially within the sentence. Punctuation marks not required to make the meaning clear or to make reading easier are being eliminated. A comparison of books printed now with those printed a century ago will usually give evidence of this development. Especially noticeable is the reduction in the number of commas.

Linguists are trying to devise a system of punctuation based on the intonations of speech. It is true that the various marks frequently serve the same purposes in print as the combinations of pause and of rising and falling pitches of voice in speech. A period marks a lowering of pitch; a question mark a raising of pitch. You can easily demonstrate this difference by reading these two sentences aloud:

The sun is shining.
Is the sun shining?

Likewise a significant pause in a spoken sentence may be marked by a comma in print.

But the marks of punctuation do not always coincide with the pattern of speech. Often they arbitrarily or mechanically used in ways not related to sound; often they indicate structural units and relationships. A discernible pause in speech may not be marked by any punctuation; for example, "He that diligently seeketh good procureth favor." Conversely, two groups of words spoken without pause may be separated by a common—usually to make the meaning clearer; for example, "What he thinks, he says." As in this example, the same words may be spoken with pause and change of pitch by one person, with none by another.

Therefore, while it may be helpful to point out some similarities in the significance of the intonations of speech and of the marks of punctuation, it seems clear that any complete description of our system of punctuation must be based on structural relationships, meanings, and usage rather than on the intonation of speech.

This section describes the principal uses of the conventional marks of punctuation, with special emphasis on divided or unsettled usages (variants).

The Period[1]

1.1 The period is almost always used to mark the end of a direct statement.

1.2 A period is used after many abbreviations, especially in nonscientific contexts; for example, *Wyo., pop. 1685, U.S.S. Decatur, Dec. 5, 220 Fifth Ave.*

Variants: Refer to the section on Abbreviations for descriptions of usages in scientific and technical contexts.

1.3 All-capital abbreviations of agencies, departments, and the like are written without periods or spacing; for example, NATO, TVA, CST (Central Standard Time), TAMU (Texas A&M University).

1.4 A period (decimal point) is used in writing decimal values; for example, 0.165.

1.5 A period is usually used after designations like "Figure 1" and "Table II" in captions; for example, "Figure 1. Schematic diagram of the circuit." But the title following the number may or may not be punctuated. *Variants*: In some journals such designations are printed with a period and a dash; in others with no punctuation at all.

1.6 Paragraph headings are usually followed by a period or by a period and a dash; for example,

> *Content of House Organs.* The material in house organs varies from company to company. . . .

1.7 Headings set in lines by themselves, whether as sideheads or centerheads, are left unpointed.

1.8 Contractions formed by omitting letters of many words are followed by periods; for example, *mfg., advt., secy.* But in contractions with omitted letters indicated by apostrophes, the period is not used; for example, *wouldn't, can't, she's.*

The Question Mark

2.1 The question mark is used after a direct question, usually indicated by a rising pitch in speech; for example, "Did you see the accident?" Though it is usually marked by inverted word order, the question may be in the form of a statement; for example, "You saw the accident?"

2.2 What appear to be questions with inverted word order may really be commands or requests ending with falling pitch and not requiring the question mark; for example, "Will you please submit a report immediately."

2.3 When a part of a sentence ending with a question mark would normally be followed by a comma or period, the comma or period is usually omitted; for example, " 'What is freedom?' he asked." and "He asked, 'What is freedom?' "

2.4 A question mark, usually in parentheses, is used to indicate that a date, a word, or a phrase is not authentic; for example, "397(?) B.C.," "His first name (William?) was not listed."

2.5 A title or heading that is a question is followed by a question mark; for example, "Shall we try for Mars?"

The Exclamation Point

3.1 The exclamation point is used to indicate forceful utterance or strong feeling; for example, "Don't do that!"

3.2 In parentheses within a statement the exclamation point may be used to mark a surprising statement; for example, "When he was 85 (!) he was still painting flagpoles."

3.3 The exclamation point is used infrequently in scientific writing.

The Comma

The comma has two common uses:

4.1 To *group* or *set off* words, phrases, or clauses. In this use a pair of commas serve as weak parentheses or dashes, and are often interchangeable with those marks, as in the following example.

The comma (from the Greek *Koptein* to cut) is used frequently.
The comma–from the Greek *Koptein* to cut–is used frequently.
The comma, from the Greek *Koptein* to cut, is used frequently.

All three sentences are correctly punctuated, but the degree of emphasis given to the interpolated phrase varies.

4.2 To *separate* words, phrases, clauses, or other items not closely related logically or grammatically. Stronger marks of separation are the period and the semicolon.

In neither of these uses is practice uniform or consistent. Especially in the grouping of words, phrases, or clauses, usage is tending away from the free "sprinkling" of commas.

Commas That Set Off

5.1 Two commas are used to set off words, phrases, or clauses that interrupt the flow of the sentence, but are too closely related to justify dashes or parentheses. Typical examples are:

Slaves, however contented, are never happy.
Scientific writing, on the other hand, is like reporting.
Adlai Stevenson, once ambassador to the United Nations, was from Illinois.

5.2 When the parenthetical expression comes at the beginning or end of a sentence, only one comma is required, whereas two curved marks would be used for a parenthesis in the same position; for example,

The lines are parallel, as shown in figure 1.
The lines are parallel (as shown in figure 1).

5.3 Loosely nonrestrictive adjective clauses are usually set off by commas; for example,

John Alsop, who has worked hard this year, will be promoted.

The relative clause describes but does not identify John Alsop. Contrast this statement with the following:

The student who studies deserves to pass.

Here the relative clause *who studies* identifies and limits the subject.

5.4 The same modifier might properly be set off in one context but not in another. "My brother, who lives in Chicago, is an engineer." "My brother who lives in Chicago is an engineer." In the first statement it is clear that the speaker has only one brother. But in the second statement the relative clause is restrictive; it indicates which of two or more brothers is the subject. The difference in intonation when the statements are spoken is indicated by the difference in punctuation.

5.5 A common fault in the use of commas to set off nonrestrictive modifiers in the middle of a sentence is to omit one of the two commas. The mistake is exactly comparable to the omission of a mark of parenthesis. It leaves the reader guessing where the interpolated element begins or ends. Compare the following:

The decision he concluded, was not easy.
The decision (he concluded) was not easy.

5.6 Transitional words and phrases are frequently not set off by commas, especially if the meaning is clear without punctuation. Here are some typical examples:

The decision therefore was a wise one.
The decision on the other hand was a wise one.

COMMAS THAT SEPARATE

6.1 Items in addresses and dates are usually separated by commas; for example, "Mr. Thomas Norton, 1365 Circle Drive, St. Louis, Missouri"; "6 P.M., October 31, 1963." (The month and day of the month are treated as one item.)

6.1.2 *Variants*: Phrases with month and year, but without the day of the month, are punctuated in three ways:

In May, 1973, he enlisted.
In May 1973 he enlisted.
In May, 1973 he enlisted.

The first usage is the oldest and most common, but the second is becoming more popular. The third is likely to be misleading since the comma may seem to separate parts of the sentence.

6.1.3 The items of a bibliographical or footnote reference are usually separated by commas; for example,

Frances Christensen, "Notes Toward a New Rhetoric," *College English*, Volume 25, No. 1, 1963, pp. 7-11.

6.1.4 *Variants*: Other punctuations used in limited areas or by certain journals are illustrated below:

(a) Items in two groups separated by a period; for example,
Rickard, Thomas Arthur, *Technical Writing*, 3rd edition. John Wiley & Sons, 1930.

(b) Items in more than two groups separated by periods; for example,
Noyes, Alfred. *The New Morning*. "Victory," p. 10. New York, Frederick A. Stokes Company, 1919.

(c) Items in groups separated by periods, with a colon after the place of publication; for example,
Updike, Daniel B. *Printing Types: Their History, Forms, and Uses*. Vols. I and II. Cambridge: Harvard University Press, 1937.

(d) Colon following the name, with commas separating other items; for example,
Osler, W.: *Modern Medicine*, ed. 3, Philadelphia, Lea and Febiger, 1927, vol. 5, p. 66.

In bibliographies surnames appear first as in a, b, c, and d and in alphabetical order; in footnotes Christian names appear first.

6.2 Two pointings are common with such expressions as *namely, viz., i.e., e.g.; for example*:

6.2.1 Setting them off by commas. This punctuation is now preferred when the term is followed by an informal enumeration, usually of words in apposition with the word preceding; for example,

We shall find two phases, namely, liquid solution and solid crystals. . . .
These were generally introduced by wax cuts, i.e., engravings made by. . . .

6.2.2 Using a semicolon before them and a comma after. This punctuation is common when a formal enumeration or a complete statement follows; for example,

The plan has four major purposes; namely, to conserve food, to provide a reserve. . . .

A compound adjective is a unit modifier; that is, it is made up of words which together modify the following noun.

6.3 Most authorities state that adjectives in a series preceding the noun they qualify should be separated by commas if the adjectives are coordinate in their application. If, however, each adjective modifies the noun as qualified by the intervening adjective or adjectives, the commas are omitted; for example,

dark, fertile loam, *but* dark sandy loam
short, swift streams, *but* short tributary streams

6.3.1 *Variants*: Occasionally the commas are omitted in such series of adjectives in scientific prose. *Suggestions to Author,* of the U.S. Geological Survey, lists these examples:

massive cross-bedded fine-grained cliff-forming dark-red sandstone
General Electric ATB 750-kilovolt-ampere 189-ampere 3-phase 60-cycle 2,300-volt
18-pole revolving-field alternating-current machine

Electrical Engineering commonly omits the commas, as in

a 100,000-kw cross-compound 3,600-rpm unit
24-kv 350,000-circular- mil section-type three-conductor cable

Civil Engineering, however, usually inserts the commas, as in

a 25-ohm, variable, coil resistance
3-phase, 60-cycle, 440-v, alternating current

6.4 In conservative scientific and technical usage the comma is usually inserted before *and* or *or* marking the final term in a series; for example,

a red, white, and blue flag
in the office, the laboratory, or the classroom

6.5 *Variant*: In general usage, particularly journalistic, the comma is regularly omitted before *and* or *or* marking the close of a series of words.

6.6 Coordinate main clauses joined by a conjunction (*and, but, for, or, nor,* and sometimes *yet*) are usually separated by a comma; for example,

The clouds above us were cirrus, but those on the horizon were cumulus.

6.6.1 *Variants*: Two variants of this usage are common:

(a) If such coordinate clauses are short and there is no possibility of misreading, the comma may be omitted; for example,
One car turned left and the other turned right.
(b) If the coordinate clauses are long and involved and have internal punctuation, the conjunction may sometimes be preceded by a semicolon; for example,
The superintendent on the job, himself an engineer, was not in his office; but his assistant, a student apprentice, gave me the information I sought.

6.7 A comma is used to prevent misreading; that is, to separate words that otherwise might be read as a single logical unit; for example,

As soon as he climbed on, the train began to move.
Ever since, the regulator has functioned perfectly.

6.8 A comma is not used to separate closely related grammatical elements; for example,

Driving on a four-lane highway with no curves and with no hills [no comma] can become monotonous.
He lived in a two-story, brick, seven-gabled [no comma] house

But if the omission of a comma might result in a misreading, the grammatically related units may be separated; for example,

Whatever is, is right

6.9 In large numerals a comma is used after digits indicating thousands; for example,

4,930, 1,256,842, 10,000

The Semicolon

The semicolon functions as a strong comma or a weak period.

7.1 The commonest use of the semicolon is to separate two grammatically independent statements not connected by a coordinating conjunction; for example,

The time was August 22, 1963; the occasion was the celebration of the author's birthday.

7.2 *Variant*: In informal, but rarely in scientific and technical, writing, such short independent statements are frequently separated by a comma.

The air was stuffy, the room was full of smoke.

7.3 Between long independent statements, especially with internal punctuation, a semicolon may be used even when they are joined by a coordinated conjunction; for example,

His younger companion, a sophomore from a Midwestern college, was big and strong; but the other, a former cowhand from Wyoming, was old and rather slight and feeble.

7.4 Semicolons are used to separate items in a series in which one or more of the items has internal punctuation; for example,

The contractors for the work were: substructure, Senior and Palmer; superstructure, except concrete floor slabs, American Bridge Company; concrete floor slabs, LaFere Grecco Contracting Company; and removal of existing substructure, General Contracting and Engineering Company.

7.5 When the second clause in a compound sentence is introduced by a conjunctive adverb (*however, nevertheless, consequently, thus,* etc.) or an adverbial phrase (*in fact, on the other hand,* etc.) a semicolon is required.

You have recommended this applicant; therefore we shall give him a trial.

7.6 A semicolon is not used after the salutation of a business letter. The proper mark is a colon.

7.7 A semicolon is not used after an introductory statement ending with "the following" or its equivalent. The proper mark is a colon.

The Colon

8.1 The colon is used as a mark of anticipation, to introduce examples, an explanation, or a quotation, often after "the following" or equivalent words.

8.2 To introduce examples:

Some of the most widely used desk dictionaries are these: *Webster's New Collegiate, Webster's New World, American College, Standard College,* and *American Heritage.*

8.3 To introduce an explanation:

It was a skillful bit of driving: he missed both cars completely.

8.4 To introduce a quotation:

The first direction in the leaflet is: "Complete the enclosed enrollment card."

8.5 To mark the close of a salutation of a letter or the opening of a speech:

Dear Sir:
Mr. Chairman:

8.6 To separate hour, minute, and second figures.

at 11:30 a.m.
2:20:5 (2 minutes 20½ seconds)

8.7 To separate chapter and verse figures in Biblical references.

I Corinthians 13:4-7

8.8 The colon is used to separate the two figures indicating a ratio, a double colon to separate two ratios:

7:10
4:5::x:20

The Dash

9.1 A dash is used to indicate a sudden change in construction or an interruption of a sentence. (For this purpose it is used infrequently in scientific writing.)

Instead of using the fire escape—but you aren't interested in what I did.

9.2 Dashes are used in place of commas or parentheses to set of interpolated words. One dash only is used if the interpolated matter is at the end of the sentence.

Two cities in this area—Detroit and Chicago—have serious integration problems.
Two cities in this area have serious integration problems—Detroit and Chicago.

In both of the examples commas or parentheses could have been used.

9.3 Sometimes a dash is used to set off appositives introduced by *namely, for example,* or *that is* (*i.e.*).

Three men were nominated for the office—namely, Ed Bowman, John Lisbon, and Bill Kent.

9.4 The dash is used frequently to indicate numerical range.

from 100—125 volts
at a 100—200% markup
between 360—390 degrees

In the first two examples the dash substitutes for "to"; in the last, for "and." The last can also be written "from 360—390 degrees." Such expressions may also be written out in full; for example,

ratings from 1 to 4 volts
from 18 to 54 inches in diameter

9.5 The dash preceded by a period is often used after a paragraph heading, thus:

Approximations.—It is no longer common to distinguish . . . .

9.6 Do not confuse the dash and the hyphen. The hyphen is used to divide compound words (self-driven) or words at line ends. The dash is a longer mark, usually made on the typewriter by doubling the hyphen (--).

The Apostrophe

10.1 The possessive singular of most nouns is formed by adding *'s;* for example,

man's, girl's, horse's, princess's, Lewis's

10.2 *Variant*: Possessives of nouns ending in *s, x,* or *z* are often formed by adding only the apostrophe; for example,

Charles', King James', Delacroix', hostess'

10.3 The possessive of plural nouns ending in *s* is formed by adding the apostrophe only; for example,

girls', horses', institutions', Joneses'

10.4 The possessive of plural nouns not ending in *s* is formed by adding the apostrophe only; for example,

men's, deer's, larvae's

10.5 The plurals of letters, figures, and symbols may be formed in two ways:

(a) By adding *'s;* for example,
3's and 6's, P's and Q's, &'s and %'s, Ph. D.'s
(b) By adding *s* without an apostrophe; for example,
3s and 5s, Ps and Qs, &s and %s, Ph. D.s

10.6 The plurals of words referred to *as words* are formed by adding *'s;* for example,

too many *and's* and *but's, damn's* and *s.o.b.'s*

10.7 The apostrophe is used to indicate the omission of letters in contractions; for example,

doesn't, we're, you're, won't (will not), he'll, '09 (1909)

Note: The possessive pronouns are written without apostrophes; for example,

yours, theirs, its, ours, hers

10.8 The apostrophe is usually retained in expressions of time and value; for example,

2 hours' retention, a dollar's worth, six weeks' period, 10 cents' worth

10.9 *Variant*: Because the idea of possession is so remote in such expressions, they are frequently written without the apostrophe; for example,

60 days notice, a thousand dollars reward, 10 cents worth

10.10 It is common to omit the apostrophe in a possessive forming part of a title or the name of a firm or organization; for example,

Governors Island, St. Elizabeths Hospital, Peoples Savings Bank

Note: The safest practice is to use the form authorized by the firm or organization.

Ellipsis Periods

11.1 To indicate an omitted word or words from a quotation, three periods are inserted; for example,

> ". . . our fathers brought forth a new nation. . . ."

The fourth period at the end of this example is the terminal period, added when the ellipsis is at the end of a sentence.[2]

Quotation Marks

12.1 Quotation marks are used to set off matter in text; for example,

> Why did Lincoln begin his speech "Four score and seven years ago" when he might have said "eighty-seven years ago"?

12.2 If the quotation is longer than a sentence or two, the passage is written with wider margins or less interlinear spacing than in preceding and following matter, and without quotation marks.

12.3 When question marks occur with periods or commas, the prevailing American practice is to place the periods and commas inside the quotation marks; for example,

> He used the phrase "posthumous impromptu."
> "Just a minute," he replied.

12.4 *Variant*: The *logical* practice of placing periods and commas outside the quotation marks if they do not belong to the quotation is preferred by such distinguished authorities as the *Oxford English Dictionary* and the *Dictionary of Modern English Usage,* by H. W. Fowler, and may be used if a writer prefers it.

Exercise

Punctuate the following sentences, and be prepared to justify your punctuation.

1. Will you kindly submit your report not later than November 19
2. But it is not important to ask What is the basic principle involved
3. Bacteria are everywhere in the air in water in milk in dust in soil in the mouth and on the hands
4. Preliminary practice was given as follows in the formboard by two trials with eyes open in the star tracing by fifteen minutes of practice
5. We have seen how the people of this state work how they earn their living now let us see how they play how they use their leisure
6. These plants are located in Newberry Cambridge Saugus Watertown and Rowley Massachusetts Portsmouth New Hampshire and Lincoln Rhode Island
7. The soil which in places overlies the hard rock of this plateau is for the most part thin and poor

8. Zinc oxide is substituted for calcium oxide and selenium and charcoal are added
9. The last two fields have much promise for the future for these fields have only recently been open for exploration
10. It begins with a discussion of different types of soil and goes on to explain the care of various kinds of flowers
11. He gestured with his arm and after pointing toward the instrument panel ran into the other room
12. South America may be divided into two sharply contrasted regions namely (1) the Pacific and (2) the Atlantic drainage basins
13. Errors in spelling typesetting punctuation or sentence construction lead to humorous statements
14. The green trees the sight of the hills and fresh raw meat had quickly revived the beast
15. A pencil placed in a glass of water appears to be bent the moon appears to be larger at the horizon than when it is overhead and lake water appears to be green or blue depending on the depth the position of the observer and other factors
16. Potatoes rice spaghetti or barley may be added for bulk and thickening
17. From the many quaint old fishing villages have come some of the most dramatic exciting legends of all time
18. Cheetahs live in various regions in Africa and Asia where they are able to find deer and antelope
19. Any tissue the dry substance of which is composed largely of proteins may be considered as protein tissue
20. The learning and staging of operettas dramatically insipid and musically void may vitiate a years work in music
21. Find the annual premium which a man born December 2 1908 must pay
22. Emporia the prophets own country with its 10000 inhabitants bought more than 2500 copies
23. The Secretary deplored the destruction of the original or virgin forests
24. The expression Where am I at is a provincialism
25. These sepals serve as a productive covering for the rest of the flower in the early stages of its development that is when it is in bud
26. Soils may be put into three groups namely clay sand and loam
27. On the sandy shores beneath fishermen spread their nets to dry
28. To neglect the integrity of the family or the prosperity of any considerable social class will sooner or later injure society as a whole
29. In 1888 Roux working in Pasteurs laboratory found that the diphtheria germ produces a toxin which causes symptoms of the disease
30. A soldier is no better than his feet is an old saying that is true for all of us
31. Mail and stagecoach lines were established traveling from St. Joseph Missouri and Atchison Kansas over the Oregon Trail to California
32. The Governor having finished his investigation steps are now being taken toward legal action
33. He maintained that if prices are lowered costs must be cut
34. He works for a good reliable old American firm
35. This method is easy that difficult
36. They argued the question at length however the Courts decision went against them

37. To say it cant be done is just an evasion
38. The technical terms rays streams jets were confusing
39. He asked quietly What do you suppose they meant by saying Havent you heard
40. They would refer to each other as my honorable opponent my rival and the gentleman of the opposition

Notes

1. The decimal system of designating the items in complicated analyses is used in this summary of usages in punctuation. See the section on Numbering and Lettering in chapter 4.
2. For further explanation of ellipsis periods refer to the chapter "Documentation."

20

Sentence Structure and Diction

In all good prose the structure of sentences conforms essentially to the same principles. Since scientific prose is usually formal—rarely slangy or colloquial—it conforms more rigidly and consistently than nonscientific prose. Therefore the author of scientific prose should be familiar with the principles.

In its choice of words too—that is, its diction—since the purpose is to convey information accurately, scientific prose must be as exact a possible.

In this chapter, however, no attempt thas been made to analyze all of the faults of scientific writing. The purpose is rather to explain some of the more common mistakes and to demonstrate how they can be avoided.

Most of the sentences used in the explanations and exercises have been lifted verbatim from papers written by students in a course in the writing of scientific papers and reports. A few examples are taken from other sources.

PARALLELISM

The principle of parallelism is that any two or more words, phrases, or subordinate clauses that are *logically* coordinate are expressed in the same grammatical form.[1] It is analogous to the mathematical principles that only similar things can be added: you can't add eggs, engines, and elephants. Neither should you coordinate nouns, verbs, and adjectives.

On the formal and colloquial levels the principle is often violated, even by careful writers and speakers. In narrative, for example, such a sentence as the following would not be rare:

> She walked through the snow, the big flakes clinging to her hair and some caught on her eyebrows.

On the formal level, this sentence would end, "some catching on her eyebrows." In scientific papers and reports, the principle of parallelism is observed by careful writers.

Ideas are logically parallel that are or can be joined by *and, or but, nor*. The parallelism is made more apparent if the coordinate constructions are numbered or lettered; for example,

> Mathematical theory (1) cannot determine and (2) is not concerned with (a) the truth or (b) the falsity of these assumptions.

It can also be demonstrated by diagraming:

```
                        { cannot determine
Mathematical theory    {        and                    { the truth
                        { is not concerned with       {     or        of
these assumptions.                                     { the falsity
```

Grammatically, five different kinds of parallelism can be distinguished.

Parallelism of Words

The parallelism of words, in pairs or in series, is so frequent and apparent that only careless writers are likely to violate the principle governing it. Typical examples, with criticisms and corrections, follow:

1. The tools required for a good interview are good health, good appearance, and a good conversationalist.

The three nouns "health," "appearance," and "conversationalist" are grammatically, but not logically, parallel. A "conversationalist" cannot be added to "health" and "appearance" any more than elephants can be added to engines. What the writer probably meant was:

> The qualities required to make a good impression in an interview are good health, good appearance, and ability to converse.

Since the notion that an interview is a job to be performed with tools is a bit far-fetched, "tools" is changed to "qualities." "Required to make a good impression in" is more exact and complete than "required for a good impression."

2. It includes definitions, descriptions, and illustrates examples of various types of scientific prose.

Here, at first reading, "and" seems to mark the last in a series of three nouns. Logically, however, it connects two verbs, "includes" and "illustrates." The sentence has *false parallelism*. Only one verb is needed:

> It includes definitions, descriptions, and other types of scientific prose.

3. At that time there were many men who had the engineering training, but were reluctant or couldn't take the executive duties.

The two predicates "were reluctant" and "couldn't take" are not parallel: one cannot write "were reluctant the executive duties." To be parallel the first predicate would have to read "were reluctant to assume."

> At that time many men with the necessary engineering training were reluctant to assume or could not assume executive responsibility.

"Many men with the necessary engineering training" is more concise than "there were many men who had the engineering training." Repeating "assume" adds emphasis, though "were reluctant to or could not assume" would be correct.

4. The report writing during the senior year will provide sufficient practice in curve plotting, contents of report, and the formal procedure involved.

The three objects of the preposition "in," though they are all noun constructions ("plotting," "contents," and "procedure"), are not logically parallel. "Provide sufficient practice in the contents of reports" is also meaningless. All three objects should be gerunds.

> The report writing during the senior year will provide sufficient practice in plotting curves, planning tables of contents [outlining reports?], and writing on the formal level.

Parallelism of Phrases

When the first of two or more coordinate ideas is expressed in a prepositional, a participial, or a gerund structure, the others should have the same grammatical form.

> 5. The base is a ½ in. by 2 in. by 3 in. cast iron block having a 2 in. by 3 in. surface as the bottom and the opposite side a supporting pivot for the arm.

"And" seems to connect the prepositional phrase "as the bottom" and the noun "side." What it logically connects is doubtful. Perhaps the meaning is best expressed without parallelism:

> The base is a cast-iron block ½ in. by 2 in. by 3 in. resting on one of its 2- by 3-in. dimensions, with the top fashioned into a supporting pivot for the arm.

For an explanation of the hyphens see the section on Hyphenation in the preceding chapter.

> 6. From the above statements it can be seen that a barometer has two uses: one, as measuring atmospheric pressures, two, as a means of measuring elevations.

Neither of the two appositives of "uses" is expressed in the most convenient form. Infinitives would be better:

> From these statements it can be seen that a barometer has two uses, to measure atmospheric pressures and to determine elevations.

In such a short sentence the use of the colon followed by "one" and "two" is unnecessary.

> 7. These ratings were low for the ease of measurement of current and voltage and ease of provision of load for the transformers, which was a load rack.

Here the writer has secured grammatical parallelism by repeating the noun "ease," but the construction is awkward and the meaning uncertain. Infinitive phrases would be better:

> These ratings were low to make it easy to measure current and voltage and to provide a load for the transformers.

The clause "which was a load rack" belongs in another sentence.

> 8. Jackscrews are used in moving houses, installing boilers, installing generators, and numerous other pieces of heavy equipment.

The last item in the series contains no gerund. Really only two gerunds are needed, the objects of the second one constituting a subordinate parallelism.

> Jackscrews are used in moving houses and installing boilers, generators, and other pieces of heavy equipment.

Parallelism of Subordinate Clauses

When the first of two or more logically coordinate ideas is expressed as a subordinate clause, the others should be in the same construction.

> 9. It is essential to know what material is important, how to arrange the parts clearly, and the most effective means of recording the facts.

Of the three objects of "to know," the first is a noun clause, the second an infinitive phrase, and the last a noun modified by a phrase. All should be clauses:

> It is essential to know what material is important, how the parts can be arranged clearly, and how the facts can be recorded effectively.

> 10. If on one film the red of the subject is recorded by making the exposure through a film transmitting red only, and another film the green by the use of a green filter, and on a third the blue filter, the three fundamental colors which singly and in various combinations comprise all the colors of the subject will be recorded.

The second and third of the conditional clauses should be constructed according to the pattern of the first, though they may be *elliptical* (that is, parts clearly understood may be omitted):

> If on one film the red of the subject is recorded by making the exposure through a filter transmitting red only, and on a second film the green through a filter transmitting green only, and on a third film the blue through a filter transmitting blue only, the three fundamental colors which singly and in various combinations comprise all the colors of the subject will be recorded.

Note that the words "of the subject is recorded by making the exposure" are understood in the second and third clauses.

> 11. These men were interested in knowing such facts as the quality of lime their soil needed, what kind, why and how often it must be replaced and what could they expect in return for their efforts.

The series of "facts" must all be expressed in the same construction. Since some of them cannot be expressed except as clauses with subjects and verbs, all should be in clause form:

> These men were interested in knowing how much and what kind of lime their soil needed, why and how often they should replace the lime, and what profit they could expect from their efforts.

Note the economy of wording effected by the secondary parallelism in the first two clauses: instead of "how much lime their soil needed and what kind of lime their soil needed" we read "how much and what kind of lime their soil needed."

Parallelism with Correlative Conjunctions

When correlative conjunctions—always used in pairs, like "either . . . or," "neither . . . nor," "both . . . and," "not only . . . but (also)," and "whether . . . or"—join two or more constructions, whether words, phrases, or clauses, the constructions must be grammatically parallel.

12. Leveling is the process by which either (I) the elevation of a given point may be found or (II) by which an unknown point of given elevation may be found.

"By which" precedes "either" but follows "or." Omitting the second "by which" secures parallelism, but the repetition of "may be found" is unnecessary. A better wording would be:

Leveling is the process by which either the elevation of a given point may be determined or a point at a certain elevation may be located.

13. The widths of the lines are of special importance in a tracing, not only to insure an easily read drawing but because there is not the contrast between wide and narrow lines in blueprints as there is in an inked drawing.

"Not only" precedes an infinitive phrase; "but" precedes a subordinate clause. In the following revision the parallelism is corrected and the awkward, obscure wording is improved:

In a tracing, the widths of the lines are of special importance, not only because the widths must be easily distinguishable to insure easy reading, but because the contrast in width of lines in blueprints is less than in an inked drawing.

14. The employee may either be reprimanded or discharged.

"Be reprimanded" is not parallel to "discharged." The sentence can be written four different ways, with resulting differences in emphasis:

The employee may be { either reprimanded
or discharged. }

The employee may { either be reprimanded
or be discharged. }

The employee { either may be reprimanded
or may be discharged. }

Either the employee may be reprimanded
or he may be discharged.

Of course such sentences may also be written with simple conjunctions.

The employee may be reprimanded or discharged.

The effect is to remove the emphasis upon the choice of alternatives.

Parallelism of Main Clauses (Balanced Sentences)

Another kind of parallelism between grammatically independent statements is often used for emphasis, usually to set off two or more ideas against each other so that their similarities and differences will be more effectively displayed. Such balanced constructions are common in oratory:

> I was born an American; I will live an American; I will die an American.

In scientific prose, however, such use of balance for emotional effect is rare. But the use of balanced constructions to bring two ideas into sharp focus together is often desirable.

> 15. Copper and iron form a thermocouple that is used for very low temperatures, while for extremely high temperatures platinum and an alloy of platinum and rhodium are used.

Here more effective balance can be secured by closer parallelism between the two statements:

> For use at low temperatures a thermocouple is made of copper and iron; for use at high temperatures it is made of platinum and an alloy of platinum and rhodium.
>
> 16. We saw electric water heaters made from start to finish; then we saw electric stoves made from start to finish.

This sentence is grammatically balanced, but it is doubtful whether the two ideas deserve the special emphasis given them. The balanced construction hardly justifies the lack of economy in wording.

> We saw electric water heaters and electric stoves made from start to finish.

* * * * *

Violations of the principle of parallelism may be due to ignorance of grammatical forms or simply to laziness. Consciously or unconsciously writers choose a certain grammatical form for the first item in a series. When they find that form unsuitable for a subsequent item, they may lazily shift to a different form without bothering to be consistent. Often a construction suitable for the first item of a series may prove to be awkward or impossible for a later item. Careful writers go back and change the form of the first item to one suitable for all items in the series.

> 17. Analysis of faculty opinions on the questionnaire indicates:
> (a) A hesitation by many to change to the semester plan because of the difficulties involved.
> (b) That many staff members who favor the quarter plan recognize the greater efficiency of the semester plan.

The noun construction chosen for (a) proved to be unsuitable for (b). The writer should have used the noun clause for (a) as well: "That many hesitate to change to the semester plan because . . . ."

To be parallel, however, phrases or clauses need not follow the same grammatical pattern, throughout. Parallel lines are not necessarily of the same length. For example, see the sentence given below.

Careless writers
and
writers who have scant knowledge of grammar } are liable to violate
the principle of parallelism.

In this example, "writers" and "writers" are parallel despite the fact that the first subject is modified by an attributive adjective and the second by a relative clause. So long as the *base* words are grammatically the same, they are parallel.

The principle applies also to items in a list, to a series of headings of the same order, to coordinate topics in an outline, and to section titles of the same series in a table of contents.

Exercises

A. *Copy from your reading in scientific prose examples of different kinds of parallelism in sentence structure. Look especially for examples in which the parallelism seems to be faulty.*
To indicate the parallelism insert numbers and letters before parallel words, phrases, and clauses, as in the following example:

In modern times there have also been (1) now and (2) again periods of great (1) freedom and (2) well-being, directed, however, (I) not so much toward (A) the intellectual life the Greeks rated so highly or (B) the artistic creation the Florentines honoured (II) as toward (A) the conquest of the material world and (B) the acquisition of wealth.—Ruth Benedict, "Who Is Superior?"

B. *Correct the faulty parallelism in each of the following sentences. If the sentences seem to have other faults, try to correct those also. Look for the faults first; then try to correct them. Physicians do not prescribe until they have diagnosed. Don't just rewrite sentences vaguely in the hope that you will hit upon an improvement.*

1. The core of this report will discuss the problem areas and how our design has dealt with them.
2. The investigation has been conducted in two phases the first being an accumulation of primary data and conglomerate financial disclosure. The second is the discussion of the data obtained in phase one.
3. The report shows what film is exposed and developed, interpretation of the finished film, and the reason for modifying the process.
4. This is a group with technical training and acquainted with procedures.
5. He likes reading books, listening to music, and he doesn't go out much.
6. Geophysics provides a method of finding oil, several kinds of ore, and the probable means of exploiting them.

7. They need to know the agreement between subject and predicate, the difference between "will" and "shall," and that to use the word *massive* three times in a single paragraph has less than massive effect.—Edward Weeks, in *The Atlantic.*
8. Methods of wind erosion control include those listed below.

    Minimizing the area of land in fallow
    Fallow land in strips among vegetated sown land
    Moderating the wind by planting shelter belts
    Cultivation of the soil when moist

9. The results of the interviews show the importance of not only good working conditions, but the importance of a good employee-supervisor relationship.
10. In the third series the reaction was present on 37 occasions, in the second it occurred 32 times, while in the third it was observed in 37 instances.
11. I propose the following modifications to laboratory practice:

    1. Formal reports to be written in the senior year only
    2. A three-hour course in technical composition during the sophomore year to replace a three hour elective
    3. Elimination of curve sheets, data sheets, wiring diagrams and complications during the sophomore and junior years.

12. Much steel is used in bracing the tank, in building the stand, and in the pipe leading from the tank to the distribution system.
13. Brown latasols may be used for raising rice, bananas, tea, or mixed farming.
14. The new design of the columns requires one new part and that two other parts must be reworked.
15. First I shall consider the points in favor of this program and second the disadvantages to the program will be considered.
16. There are many welfare agencies which provide social welfare, aid for handicapped children, and juvenile delinquents.
17. Either you must grant his request or incur his ill will.
18. This implied not only reductions in office personnel, but among enlisted men as well.
19. My objections are, first, the injustice of the measure; second, that it is unconstitutional.
20. The United States has not and will not shirk this responsibility.

AGREEMENT OF SUBJECT AND VERB

It is a simple and invariable rule that a verb must agree with its subject in number. In most sentences we apply the rule without having to reflect. But persons who are careless or who have scant knowledge of grammar are likely to violate the principle where the relationship between subject and verb is not immediately obvious.

Seven different types of sentences, as indicated by the following headings, may require conscious checking to insure that the subjects and verbs agree in number.

Subject Separated from Verb

If a subject is separated from its verb by an intervening phrase or clause, lack of agreement may result.

1. Due to war conditions abroad, information concerning new discoveries in this field have been somewhat limited.

Between the singular subject "information" and its verb intervenes a phrase with the plural noun "discoveries" which has—as grammarians explain—attracted the verb into the plural form. With this fault corrected, the sentence reads:

Because of war conditions abroad, information concerning new discoveries in this field has been somewhat limited.

"Because of" is substituted for "due to" because "due to" is in doubtful standing as a preposition introducing an adverbial modifier. The *American Heritage Dictionary* reports that 85% of its Usage Panel do not approve it in writing.

2. His reputation as well as his fortune were at stake.

Regarding the subject here as plural (equivalent to "his reputation *and* his fortune") is common in informal usage, but not yet established usage on the formal level. The singular noun "reputation" is the grammatical subject.

His reputation as well as his fortune was at stake.

Compound Subject

When the subject includes two or more separate and different items joined by "and," it is plural, and the verb should be plural. Only when the items are merely different names for the same thing will be the verb be singular. Thus one writes "The president and the chairman of the board is . . ." if the president is also the chairman; "The president and the chairman of the board are . . ." if the president and the chairman are two people.

1. Only one speed forward and one reverse is needed.

The "one speed forward" and "one reverse" are separate speeds or gears and form a plural subject. In the following revision subject and verb agree, and parallelism is improved.

Only one forward speed and one reverse speed are needed.

If the subject includes items joined by "or," the verb agrees with the item nearest to it. Thus it is correct to write. "The president or his secretary is . . ." and "The president or his secretaries are . . . ." It is also correct to write "He or I am in charge" since the verb agrees with the nearest subject "I" both in number and in person. Fortunately such odd-sounding constructions are infrequent in scientific writing.

Subject After the Verb

The rule of agreement applies whether the subject precedes or follows the verb. When the subject follows, the writer must look ahead to see whether it will be singular or plural.

1. In the committee expense budget is recorded the date of the requisition, the amount of the purchase, and the balance left to be spent.

Reversing the order of subject and verb makes clear the need for a plural verb: "the date, the amount, and the balance" clearly add up to three items.

In the committee's expense budget are recorded the date of the requisition, the amount of the purchase, and the balance left to be spent.

"Committee's" is better than "committee": don't abuse the privilege of using a noun as an adjective.

Collective Nouns as Subjects

When the subject of a clause designates a group of objects or persons or acts, the number of the verb may be singular or plural. If the group is a unit, it is singular; if it is several individual things or persons, it is plural. Both of the following statements are therefore correct, though the verb in the first is singular and that in the second is plural:

The committee meets tomorrow.
The committee are not in agreement.

Some other common collective nouns are:

army	board	crew	firm	number
association	class	crowd	group	public
athletics	company	dozen	herd	remainder
audience	contents	faculty	jury	staff
band	council	family	majority	team

Nouns with Foreign Plurals

Words from foreign languages may retain their original plural forms, may have an anglicized plural form, or may have both foreign and anglicized plural forms. One unfamiliar with these endings is likely to make such mistakes as the following:

These phenomenon [phenomena] are unusual.
Innumerable larva [larvae] had hatched there.
This strata [stratum] was concealed by soil.

Sometimes the foreign and anglicized plural forms have different meanings, as do *addenda* (items added) and *addendums* (parts of the tooth of a gear wheel).

In the following list the anglicized plural endings (when they are in use) are placed after the singular forms, the original foreign plural endings last. When listed, both forms are correct, but the anglicized plurals are becoming increasingly common in scientific writing.

agenda -das -da
alumna (feminine) -nae
alumnus (masculine) -ni
amoeba (or ameba) -bas -bae
analysis -ses
apparatus -tuses -tus
appendix -dixes -dices
automaton -tons -ta
axis axes
bacillus -li
basis bases
cactus -tuses -ti
crisis crises
criterion -ions -ia
curriculum -lums -la
datum data
diagnosis -ses
erratum -ta
focus -cuses -ci
formula -las -lae
fungus -guses -gi
genus -uses -era
hypothesis -ses
index -dexes -dices
insigne -nia
larva -vas -vae
locus loci
matrix -trixes -trices
medium -diums -dia
memoranda -dums -da
minutia -tiae
momentum -tums -ta
moratorium -iums -ia
nebula -las -lae
neurosis -ses
nucleus -cleuses -clei
opus opuses opera
ovum ova
parenthesis -ses
psychosis -ses
radius radiuses radii
rostrum -trums -tra
species species
stadium -diums -dia
stimulus -li
stratum -tums -ta
syllabus -buses -bi
synopsis -ses
synthesis -ses
tableau -eaus -eaux
terminus -nuses -ni
thesis theses
ultimatum -tums -ta
vacuum -ums -ua
vertebra -bras -brae
vertex -exes -ices
vortex -exes -ices

Indefinite Pronouns as Subjects

The following indefinite pronouns are always singular:

anybody	each	everything	nobody	someone
anyone	either	much	one	something
anything	everybody	neither	somebody	

None, however, may be either singular or plural.

Relative Pronouns as Subjects

The number of a relative-pronoun subject is determined by its antecedent and in turn governs the number of its verb. Agreement is likely to be faulty when the antecedent of the pronoun is not immediately evident.

It was one of those accidents which seems unavoidable.

The antecedent of "which" is "accidents," not "one"; therefore the verb in the relative clause should be plural.

It was one of those accidents which seem unavoidable.

Exercise

Correct the faults in agreement in the following sentences, and be prepared to justify your corrections. If you find faults other than failures of agreement, correct those also. Make no change unless you have a good reason for it.

1. At present there is no definite criteria as to what constitutes a low-pay phone.
2. Ten ml of EDTA, 15 ml of buffer solution, and a drop of the fluorescent solution is added to each aliquot.
3. This dial plate is one of several subassemblies that goes into the final assembly in making the 3½" pressure gauge.
4. Frost damage to pavements and their subgrades occur in two distinct phases.
5. My leisure was complex, and my emoluments large.
6. The reputation of the dealer as well as the written guarantee are important to many people.
7. A set of laboratory instructions are distributed during the first period of each laboratory course.
8. The results of this test shows that the hardness and tensile strength has been increased considerably.
9. From this meager beginning has developed the scholastic, professional, and Greek letter fraternities that we know today.
10. Neither David or Amy are at home.
11. The sheriff with all his men were at the door.
12. There occur in the spectrum changes which are quite exceptional in respect to the rapidity of their advent.
13. One of the ablest men who has attacked this problem has been a forester.
14. Each boy's personal disposition and problem is quickly described.
15. The complexity of the building types needed today are a reflection of the complexity of our society.
16. Into each missile goes tens of thousands of parts, each of which must work perfectly.
17. The article is focused on program content in television, a topic in which there appears to be as many views as there are people.
18. There is brisk action, dramatic Indian fights, much ruffling of young cavaliers, and a fine aristocratic swagger.
19. Observation of general functioning and mechanical adequacy of the latest design features were made.
20. Of special interest to the student is the suggested visual aids which can make the presentation more vivid and dramatic.

REFERENCE OF PRONOUNS

Since a pronoun usually stands for a preceding word or group of words (the *antecedent*), the meaning is clear only when the reader knows what the antecedent is. A careless writer, knowing what the pronouns stand for, may forget that the reader's only way of determining the antecedents is from the construction of sentences. In scientific writing especially, the reference of every pronoun should be so clear that the reader identifies it without conscious thought.

Indefinite pronouns have no antecedents; with them there is no problem. "*Anyone* can come." "*Everyone* knows that *it* is raining." "*They* say that *no one* can enter."

A common fault is to write a sentence so that a pronoun may have two or even three possible antecedents.

> 1. The combustion of the vapors causes an enormous force upon the piston, which forces it down, which in turn forces the crankshaft to rotate with an increased speed.

In the clause "which forces it down," either "which" or "it" may refer to "combustion," "force," or "piston." The second "which" has no reference except vaguely to "forces it down." The whole sentence must be revised:

> The enormous force exerted by the burning vapors forces down the piston, which in turn rotates the crankshaft.

In the revision "which" refers clearly to "piston," the noun immediately preceding it.

> If the hole isn't near the highway, it must be snaked into position with a team of horses.

The real antecedent of "it" is neither "hole" nor "highway," but "pole," a word not in the sentence. Such an apparently ludicrous statement may result from failure to clarify the reference of a pronoun. Substituting the real antecedent corrects the fault:

> If the hole isn't near the highway, the pole must be snaked into position with a team of horses.

"This" and "which" should not be used to refer vaguely to previous statements. If the antecedent is a clause and not a single word, the reference may not be immediately clear. In the following sentence "this" refers vaguely not only to the preceding clause but to preceding sentences:

> 3. There is a circular dial about one third of the way from the center to the tracing end, and all of this must of course have been constructed very carefully.

Since the statement has two distinct parts, divide it into two sentences and substitute a noun for "this":

> One third of the way from the center to the tracing end is a dial. The whole instrument must of course have been constructed carefully.

> 4. The lines must be inked in with a ruling pen, which requires a great deal of care.

The writer means that the inking, not the pen, requires care:

> The lines must then be inked in with a ruling pen, a task which requires great care.

Exercise

Revise the following sentences so that the reference of pronouns is unmistakable. If you find errors other than faulty reference, correct those also. Don't try to revise until you know what the faults are.

1. Dr. Judson is studying the phenomena to determine its significance in the field of genetics.
2. Printed-circuit theory was developed in England about 1936, but it was not until World War II that their importance and use was realized.
3. The company is glad to have engineering seniors visit their plant.
4. *Time lost due to equipment breakdown.*—This was by far the cause of the most lost time.
5. The valve is set to maintain a temperature leaving the evaporator which is above the saturation of the refrigerator.
6. Pinning may be partly prevented by use of chalk, oil, or turpentine. The latter is best.
7. Child development workers start with infants and follow their lives closely, until they become aware that they can do no more.
8. I sincerely hope that these data that I have taken the trouble to gather are of some help to you. You can rely on its authenticity of source.
9. If the die is cleaned at the first appearance of oxidation, it will be removed completely and the useful life of the die prolonged.
10. He noticed a large stain in the rug that was right in the center.
11. He wrote three articles about his adventures in the Near East, which were published in *Holiday*.
12. Secretary of State Dulles told Foreign Secretary Eden that he had a "bad habit"—doodling.
13. The broad-spectrum antibiotics have so simplified pneumonia therapy that it is now frequently treated at home.
14. The Premier agreed that France was "the sick man of Europe," but he denied she was decadent.
15. There is a city-owned pier running out from this land which is used by a marine repair firm.
16. We found a pile of picture frames in the corner, which had not been dusted for years.
17. After weighing, the sacks must be carried a considerable distance which takes considerable time.
18. Each time the trigger is pulled or the hammer is drawn back, the cylinder revolves clockwise from one hole to the next. This is operated by a small lever engaging the cylinder from below.
19. When the company is authorized by the employees, they are to deduct union dues from their checks.
20. The crew was shorthanded, so they used midshipmen to make up the difference.

DANGLING AND TRAILING MODIFIERS

Two of the most common faults in scientific writing are dangling and trailing modifiers.

Dangling Modifiers

A modifier dangles when the word that it logically modifies is omitted from the sentence, or when it seems to refer to some other word.

Dangling modifiers are usually participial, gerund, or infinitive phrases or elliptical clauses. Most frequently they occur at the beginnings of sentences.

1. Before going to work the truck is taken to the warehouse, where the supplies are replenished.

The dangling construction is "before going to work," which logically modifies "the men" or "the laborers," words not in the sentence. Since you naturally expect a modifier at the beginning of the sentence to refer to the subject of the sentence, you are led here to think of the truck going to work. But trucks do not go to work; they are driven to work.

You can correct such dangling modifiers in two ways:

Method I: Make the subject of the main clause the name of the person or thing to which the dangling construction logically refers.
Method II: Expand the dangling phrase or elliptical clause to a complete clause.

Sometimes either of these methods of correction will be feasible; sometimes only one. In the sentence given, only the first method is easily applicable:

Before going to work, the men take the truck to the warehouse and replenish the supplies.

2. When used as an altimeter, aneroid barometer readings are approximate.

Here the elliptical clause "when used as an altimeter" refers logically to "aneroid barometer," which, though contained in the sentence, is used as an adjective and not as the subject of the main clause. Hence the elliptical clause seems to modify "readings." Applying either of the two methods will correct the fault:

Method I: When using an altimeter, the aneroid barometer gives only approximate readings.
Method II: When the aneroid barometer is used as an altimeter, it gives only approximate readings.

Dangling constructions are most frequently participial phrases at the beginnings of sentences:

3. Based on this analysis and other factors, the engineer makes a decision to follow alternative X or Y.

"Based" modifies "decision," not "engineer." Applying Method I results in this sentence:

Based on this analysis and other factors, the decision of the engineer is to follow alternative X or Y.

Applying Method II results in something like the following:

If the engineer depends on this analysis and other factors, he follows alternative X or Y.

When you use the passive voice in the description of a process, you must be especially careful to avoid dangling constructions. Agents or "doers" are not usually specified when the passive voice is used; consequently a participle referring to them will usually dangle.

4. Using a wrench, the plug at the side of the transmission gear housing is removed.

If the passive voice is retained, the active participle must be dropped:

With a wrench, the plug at the side of the transmission gear housing is removed.

Also, when using the passive voice, avoid prepositional phrases that contain active gerunds:

5. After removing the crucibles from the oven, they are placed in a desiccator to cool.

The active gerund "removing" refers to the person who is performing the process, but who is not named in the sentence. If the passive voice is retained in the main clause, Method II must be used to correct the dangling construction:

After the crucibles have been removed from the oven, they are placed in a desiccator to cool.

Such sentences are common in technical writing, but you should avoid them, not only because they contain dangling constructions but also because there is a shift in point of view from the active to the passive. Here is an example from the journal *Electrical Engineering*:

6. In measuring the carrier transmission losses, some interesting loss characteristics were recorded.

If the passive voice is used, the sentence should read:

When the carrier transmission losses were noted, some interesting loss characteristics were recorded.

In passages with many mathematical equations, dangling modifiers are likely to occur:

7. Substituting the value of x from equation 1, an expression for dz/dt is obtained of the form

$$\frac{dz}{dt} = \frac{-z}{V} C + D$$

"Substituting" dangles. Both methods may be used to correct the mistake:

Method I: Substituting the value of x from equation 1, we obtain an expression for dz/dt, etc.

Method II: When the value of x is substituted from equation 1, an expression for dz/dt is obtained, etc.

Exceptions to the Rule

A limited group of words that were once participles have come to be used absolutely as sentence modifiers and no longer need the prop of a noun or pronoun to depend on. Examples are phrases beginning with *considering, assuming, speaking, taking,* and *remembering.*

8. Generally speaking, there is no objection to this procedure.
9. Assuming this hypothesis to be true, the conclusions are justified.

In these sentences there is little danger of mistaking the meaning, though such constructions are usually unnecessary. Compare these statements with the following:

In general, there is no objection to this procedure.
If this hypothesis is true, the conclusions are justifiable.

When a gerund phrase follows a main clause, the "dangler" may often be justifiable, if not unavoidable:

10. The mistake may be avoided by using another construction.

"Using" dangles because it refers to an agent ("the writer," "one") not mentioned in the sentence. Since the main verb is passive and the gerund is active, there is also a shift of voice in the sentence. But usage justifies the construction, since there is no convenient substitute. It is clear, and when the passive voice is used in the main clause, the dangling gerund can be avoided only by using less idiomatic substitutes:

The dangling gerund can be avoided only by the use of less idiomatic substitutes.
The dangling gerund can be avoided if another construction is used.

The active voice can be used throughout, of course:

The writer can avoid the dangling gerund by using another construction.

Now "using" refers to "writer."

Trailing Participial Phrases

A participial phrase "trails" when it hangs loosely and weakly at the end of a sentence. Even though the meaning may be clear, the trailing phrase weakens the sentence ending. Often it embodies a thought that deserves to be expressed as a main clause.

1. The losses are dissipated as heat, making good insulation and proper cooling a serious problem in the larger transformers.

"Making" refers neither to "losses" nor to "heat," but to the whole main clause. The construction is objectionable both because it dangles and because its content needs more emphasis than the subordinate phrase gives it. Two revisions are possible:

> Since the losses are dissipated as heat, insulation and cooling are serious problems in the larger transformers.
>
> The fact that the losses are dissipated as heat makes insulation and cooling difficult in the larger transformers.

Sentences ending with "thus," "thereby," and "therefore" followed by a participle almost invariably can be improved:

> 2. The organic material content of this soil was very low, thereby eliminating any trouble from this source.
>
> 3. The work was started early in the season, thus avoiding bad weather.

Revisions eliminating the dangling and trailing phrases might be the following:

> Since the soil contained little organic material, all trouble due to the presence of such matter was eliminated.
>
> Since the work was started early in the season, bad weather was avoided.

In both revisions the trailing participial phrase has become the main clause.

Exercise

Correct the dangling and trailing constructions in the following sentences.

1. Realizing that people don't always want the quantity that is packaged, any customer is allowed to break a package to get the desired amount.
2. The possibility for gross abuse has always existed, thus recognizing the need for the Securities and Exchange Commission.
3. The cabinet for the instrument was 18 inches long, measuring it on the inside.
4. Before submitting the final report, the features of each location will be written up and compared with each other.
5. All calculations were made using a computer.
6. By making home service calls, operating a recipe testing and developing division and giving demonstrations, customers come to know and understand the company's product.
7. Having found the cause of the trouble, the sparkplug was replaced by a new one.
8. While using the Jet Eraser, it is held in the hand in the same manner that a pencil would be held.
9. His discharge, after serving the company for many years, came as a shock to him.
10. Walking slowly down the road, he saw a woman accompanied by two children.
11. Young and inexperienced, the task seemed easy to me.
12. As a mother of five, with another on the way, my ironing board is always up.
13. Unbeaten thus far this year, the victory was his seventh in a row and the tenth since last dropping a decision last September.
14. When wiring houses, the code must be followed.
15. After adjusting the valves, the engine developed more power.

16. Transmitting motion from one part of the machine to another, the adjustment of the gears was vital.
17. After connecting this lead to pin 1 of the second tube, the other lead is connected to pin 2.
18. After drying for three days under the hot sun, workers again spray the concrete with water.
19. Enclosed herewith is a list of important essentials that should be subject to coverage in the next conference dealing with the matter of absenteeism.
20. Turning now to panel CDFG, a compressive force will be required in DG so that X_2 will appear as in Figure 3.

MISPLACED MODIFIERS

The elements of a sentence must be so arranged that the reference of modifiers is unmistakable. As a general rule, modifiers should be placed as close as possible to words modified; whether before or after depends on the modifier and on the emphasis desired. Usage permits a considerable variety of arrangement. For example, here is a simple sentence arranged in four different ways:

1. Unharmed, the captain, with his uniform all dirty, dived ingloriously into the shellhole.
2. With his uniform all dirty, the captain, unharmed, ingloriously dived into the shellhole.
3. Ingloriously the unharmed captain, with his uniform all dirty, dived into the shellhole.
4. Into the shellhole ingloriously dived the captain, unharmed, with his uniform all dirty.

In each sentence the modifiers are placed correctly. The differences are differences in emphasis.

Both adjective and adverb modifiers frequently come at the beginning of sentences:

Adjective modifier first: Unaware that the current had been turned on, the lineman touched the wire.
Adverb modifier first: With great caution, she eased the car into gear.

Such variations from the normal sentence order—subject, verb, complement—make writing more readable, but must be checked carefully to avoid misplaced modifiers.

Usually the adverb "only" should be placed before the word it modifies.

5. I think that technical language only should be used when there is no way of expressing the thought in simple English.

The writer does not mean to say that when simple English is inadequate, technical language must be used. The adverb "only" modifies "when there is . . . English" and not "technical language." The writer intended to say:

I think that technical language should be used only when there is no way of expressing the thought in simple English.

A writer sometimes misplaces a modifier because of failure to review the sentence from the point of view of the reader. What seems like an obvious relationship to the writer may be obscure to the reader.

6. He was ordered to close the switch at least three times before he obeyed.

"At least three times" seems to modify "to close"; logically it modifies "was ordered." The sentence must be rearranged:

Before he obeyed, he was ordered at least three times to close the switch.

Exercise

Revise the following sentences so that the references of all modifiers are unmistakable:

1. I have been trying to place him under contract to work here for three years.
2. This mixture, after being thoroughly blended, is transported to the site where it is to be applied in wheelbarrows.
3. Submit cards each Monday following the close of a week's work prior to 10 o'clock.
4. The new facilities will make it possible for babies to be born in the city hospital for the first time.
5. The chairman said he hoped all members would give generously to the needs of the party at a meeting of the committee last night.
6. We only found two mistakes in the report.
7. You can call your mother in London and tell her all about George's taking you out to dinner for just sixty cents.
8. The development and assay of the antibiotics by microorganisms has had a profound effect in the medical field to cite one example.
9. The following passage is quoted from a textbook which proves the point.
10. Today the Dean wishes only to see freshmen.
11. Often the fire is put out before much damage has been done by the firemen.
12. Level off the cup that has been filled with shortening with a knife or any similar tool.
13. The writer says that he intended to give up his job in the first paragraph.
14. The group will only study those problems that require lengthy investigations.
15. Accidents are frequent in machine shops and sometimes even machinists get killed.
16. The tug which was whistling noisily chugged up the river.
17. There must be no noise whatever while I am reading to distract my mind.
18. She did not notice that some of the bolts were missing until she counted them.
19. The aide was instructed to open the valve at least three times before leaving the room.
20. The application of engineering principles and theories can be applied through actual experience to industrial needs before one has graduated from college.

EXCESSIVE PREDICATION

Primer style, which consists of expressing every idea as an independent statement, is the mark of an immature mind. The sentences below are examples.

I see the cat. The cat is black. The cat's tail is long. The cat has whiskers.

Children often speak and write in such simple predications.

Only slightly more advanced than primer style is that in which independent statements are connected by *and, but,* and *so*:

> I was going to school, but I saw a big dog, and it barked at me, and I was scared, so I ran home.

The mature mind functions differently. It expresses main ideas in principal clauses; it subordinates the incidental and less important ideas.

> 1. The other piece is 5/16 in. thick. It has a rectangular shape. It is 5 3/4 in. long and 1 3/4 in. wide. It is made from a plastic material.

Of the several predications concerning "the other piece," the one concerning material is probably most important, those concerning size and shape are incidental.

> The other piece, rectangular in shape, with dimensions 5/16 by 1 3/4 by 5 3/4 inches, is made of plastic.

In an isolated sentence it is difficult to tell which fact deserves most emphasis. If the predication concerning shape is of more importance, the sentence should read:

> The other piece, made of plastic, is rectangular in shape, with dimensions 5/16 by 1 3/4 by 5 3/4 inches.

Excessive use of coordinating conjunctions (*and, but, so*) is especially undesirable when the ideas so connected are not logically coordinate.

> 2. In winter a worker may face a casting that is red-hot and his back may be chilled to the bone.

The implication is that the chill in a worker's backbone is the result of facing the red-hot casting. Perhaps the meaning intended was

> In winter while a worker's face is scorched by the heat of a red-hot casting, his back may be chilled to the bone.

Here is another example of primer style from a student's writing:

> 3. The ripe seeds fall to the ground in the fall, but they have hard coats and they do not take up any water, so they do not begin to grow until the next spring.

Note the four main clauses joined by *but, and,* and *so*. Several revisions are possible; for example,

> Because the seeds are hard-coated and do not absorb water when they fall to the ground in the fall, they do not begin to grow until the next spring.

Excessive subordination, however, is as bad as excessive predication; it may result in such complexity of structure that the reader's mind is quickly fatigued in the effort to understand it. Sentence lengths should vary. Important ideas should occasionally be expressed as simple sentences, without qualifying modifiers. Two closely related ideas of equal importance may need to be jointed by *and* in a compound sentence.

You would find it worthwhile to analyze the structure of sentences in your own writing, and try to avoid both excessive predication and excessive subordination.

Exercise

Revise the following sentences to eliminate excessive predication. Decide which of the ideas deserves most emphasis and embody it in the main clause; express minor ideas in suitable subordinate construction. Correct any errors in spelling and grammar.

1. The remaining time for this project will be spent in organizing my information into a formal report which will be submitted on May 5, 19.... Also some time will be spent in locating available facilities to rent for housing of Acme's vehicles. Approximately one week will be spent at Acme with their transportation department completing my present figures.
2. The level is the main and most important part of the instrument, and is made up of telescope, spirit level, and adjusting screws and plates.
3. Various degrees of hardness are necessary in metals, and the degree of hardness is determined upon what the metal is used for.
4. A thermostat is an instrument which is used for controlling measured temperatures and is in the general classification of a bimetallic thermometer.
5. The pivot is a small steel rod ⅛ inch in diameter and ¼ inch long. The pivot should have a small head at each end.
6. The motion is accompanied by friction to obtain heat. This friction is referred to commonly as the resistance of the wire.
7. The die is fastened to a larger piece of steel suitable for clamping to the bed of the press. This piece of steel is called a die shoe.
8. The pyrometer requires auxiliary apparatus and often some skill in reading, so its use is somewhat limited.
9. The advantages are quite numerous and a few of them are listed below.
10. The transformer is not a perfect machine so there must be some allowance made for the losses that occur in the transformer itself.
11. The upper third of the shaft is reserved for the handle. This part is therefore tightly covered with leather. The leather has two purposes. The size of the handle is increased and the latter is therefore easier to hold. One can also grip the handle more firmly than he can the metal rod.
12. Each atom of every substance is made up of a nucleus. This nucleus consists of a mass of protons and a few electrons.
13. The basement floor is cement. It contains some cracks. It drains water well to the sewer.
14. To these slots are attached the rotor blades. The blades are the working part of the turbine. They are without exception made of low carbon chrome iron alloy.
15. The objectives of the course are discussed. One of these objectives is to increase skill in the organization of scientific data. Another objective is to teach the student the difference between facts and opinions. A third objective is to increase skill in scientific writing. The fourth is to increase knowledge of the mechanics of scientific writing.
16. The front bearing of the generator needs replacing, and the commutator is in good condition, but can never be turned down again.

17. There are two rest rooms. These contain one wash basin and two stools each. The lighting in the rooms is very poor. They are both very dirty and dismal. These seem grossly inadequate.
18. The lug wrench is taken and the lug nuts are further tightened. This process prevents the lug nuts from vibrating loose when the car is being driven.
19. The stone prongs are pieces of gold which are glued on the body of the earring. There are four of these prongs.
20. There are two types of focal plane shutters. First is where the screen or shutter has a given size opening in it. The second is where the screen has a variable size opening in it.

WORDINESS

The use of ten words to express an idea when five can do it is undesirable in any kind of prose, but particularly in scientific prose. Yet even skilled writers frequently use too many words. It is much easier and quicker to write vaguely about an idea than to express it in exactly the right words. The Roman naturalist Pliny once concluded a letter, "I had not time to write you a short letter; therefore I have written you a long one."

Frequently the wordiness is the result of an inadequate vocabulary or of failure to hunt for the one word that may substitute for several. The value of technical terms is that they express a complex idea succinctly. Writers with large vocabularies can—though unfortunately sometimes don't—write concisely.

1. The device which regulates the speed of the engine makes use of the principle of centrifugal motion.

Here the writer economizes at the end of the sentence by using the term "principle of centrifugal motion," but spends words extravagantly at the beginning.

The governor utilizes the principle of centrifugal motion.

Seventeen words are condensed into eight without change of meaning. Of course, if the reader doesn't know what a "governor" is, the saving is not justified. Clarity is more important than conciseness.

Phrases or appositives can frequently be substituted economically for clauses:

2. Something that is easy to make and yet that is very decorative and beautiful is a finger ring made of stainless steel or monel metal.

In the second example change the two adjective clauses to adjectives; omit "very" (most "very's" are mere dead weight); delete "beautiful" because "decorative" is adequate; and omit the redundant participle."

Something easy to make and yet decorative is a finger ring of stainless steel or monel metal.

The revision saves eight words.

Repetition resulting from mere carelessness or laziness is inexcusable:

3. There are many ways in which a student may profit from writing a good laboratory report. I will discuss four ways in which the writing of good reports will aid a student.

One sentence can easily express the whole thought and save twelve words:

Of the many ways in which a student may profit from writing a good laboratory report, I shall discuss four.

Sentences beginning "There is . . .," "There have been . . .," etc. should be scrutinized carefully. They are often wordy and at the same time weak in that a position of emphasis—the beginning—is taken up by meaningless words.

4. There have been elaborate plans made.
5. There were four more cases of polio in the city reported yesterday.

Revise these by beginning with the subject of the predication, not the meaningless expletive:

Elaborate plans have been made.
Four more cases of polio were reported in the city yesterday.

This criticism does not apply to such sentences as "There is no hope" and "There are no fools like old fools," where dropping of the expletive would change the meaning.

Sometimes the kind of language called "jargon" will conceal a simple idea in a thick fog of many words, using circumlocutions instead of going straight to the point, preferring abstract nouns to concrete ones, concentrating on sound rather than sense. Its identifying marks are the *cliché* (the overused word), the ready-made phrase, and the abstract noun, of which the following are illustrations:

Clichés	*Ready-made Phrases*	*Abstract Nouns*
tiller of the soil	due to the fact that	character
institution of higher learning	according as to whether	case
burning the midnight oil	as far as that is concerned	nature
gentle reader	along that line	proposition
marts of trade	with regard to	occurrence
imposing structure	in connection with	situation
colorful spectacle	in that case	degree
point with pride	there can be no doubt that	standpoint
veritable mine of information	at the present moment	aspect

All of these faults of style have been pointed out and made infamous by Sir Arthur Quiller-Couch in an essay entitled "On Jargon."

Though the following sentence reads smoothly and is grammatically correct, it is mostly jargon:

> 6. The reason for the failure is due primarily to the fact that the student's fundamental interest in his work has been killed by burying his initiative beneath a constant repetition of uninteresting form.

One can only guess at the meaning, but perhaps something like the following was intended:

> The student fails because his interest has been killed by repetition of an uninteresting form.

The meaning is expressed in 15 words instead of 33.

Using too many words to express an idea frequently results in obscuring the thought. The wordy sentence is like a picture taken with the lens out of focus: the image (thought) is dim and fuzzy, not clear and distinct.

> 7. Although the losses in the power transformer are very small, there are some, and they constitute about 2 percent of the input at rated load.

Two main clauses and a subordinate clause are used to express a single idea. It can be expressed in a single predication, with a saving of eight words:

> The losses in the power transformer are only about 2 percent of the input at rated load.
>
> 8. In order to find the water equivalent of the copper vessel, the specific heat of the copper must be looked up and you must multiply this number times the mass of the vessel in order to find the water equivalent.

Notice the unnecessary repetition and the confusing shifts of construction. Elimination of the deadwood reduces the length of the sentence from 45 words to 27:

> The water equivalent of the copper vessel is the product of the specific heat of the copper, the change in temperature, and the mass of the vessel.

Exercise

Condense the thought of the following sentences into as few words as you can without changing the meaning. Correct any errors in spelling and grammar.

1. It should be mentioned that in the case of several observations there is room for considerable doubt concerning the correctness of the dates on which they were made.
2. There are two cases where sunspots changed with considerable rapidity.
3. In three cases the stars are red in color.
4. In my humble opinion, though I do not claim to be an expert on this complicated subject, fast driving, in most circumstances would seem to be rather dangerous in many respects, or at least so it would seem to me.
5. The average doctor who practices in small towns or in the country must toil day and night to heal the sick.

6. When I was a little girl I suffered from shyness and embarrassment in the presence of others.
7. It is absolutely necessary for the person employed as a marine fireman to give the matter of steam pressure his undivided attention at all times.
8. There can be no doubt that many institutions of higher learning boast of buildings that are imposing structures.
9. A period of unfavorable weather set in.
10. The question as to whether he was guilty remains an open one.
11. Colorado is a state that attracts visitors because of its skiing facilities.
12. Highway deaths are down eleven percent in the state over the period a year ago.
13. The supervision of driver and safety education at the state and local levels should be assigned to personnel qualified by virtue of their adequate social characteristics and specialized training and experience in the field.
14. Greater success has been enjoyed this year than last in the case of the engineering department.
15. From a cleaning point of view, these valves are relatively good.
16. A person can take a bucket of water and swing it around in a circle and if the speed is great enough no water will spill.
17. T & AM 327 may be regarded as somewhere between a formal and informal course so far as reports are concerned.
18. There is an experiment which is often performed in the a-c laboratory and which illlustrates this point.
19. Indeed, there may be some experiments that are of little value to the student.
20. There are many characteristics in common for the systems mentioned above.

CLEARNESS

The first requirement for good scientific writing is that it should be clear. "Our words should fit our thoughts like a glove and be neither too wide nor too tight," wrote George Herbert Palmer. "If too wide, they will include much vacuity beside the intended matter. . . . Too frequently words signify nothing in particular. They are merely thrown out in a certain direction to report a vague and undetermined meaning or even a general emotion."

To write exactly is not easy. It requires constant attention to the meaning of words and to their arrangement in the sentence. Language at its best is an imperfect medium of communication. To use it exactly requires, as Robert Louis Stevenson said in "Truth of Intercourse," "a perpetual determination not to tell lies; for of course every inaccuracy is a bit of untruthfulness."

Lack of clarity may result from violations of any of the principles of sentence structure and diction already studied. In this section, however, three additional sources of obscurity are analyzed:

1. Misused words
2. Illogical sentence structure
3. Overuse of technical terms

Misused Words

Careless or illiterate people sometimes confuse the meanings of words.

1. The diameters of the rods are ¼ inch, ⅜ inch, and ½ inch respectfully.

"Respectfully," denoting deference or respect, is confused with "respectively," meaning "in the order given."

2. The wind has considerable affect upon tide levels.

"Affect" is improperly used as a noun except as a technical term in psychology.

3. I was made a full-pledged brakeman on the Milwaukee railroad.

"Pledged" (for "fledged") is an example of the "malaprop" or "malapropism," a grotesque misuse of words resulting from an ignorance of their meanings.

The following list contains some of the pairs and groups of words frequently confused in scientific writing:

ability, capacity
accept, except
adopt, adapt
affect, effect
already, all ready
altogether, all together
alumnus (-i), alumna (-ae)
among, between
amount, number
apt, liable, likely
attain, retain
balance, remainder
center, middle
contain, hold
continual, continuous
credible, creditable, credulous
discover, invent
equable, equitable
explicit, implicit
farther, further
fewer, less
good, well
healthful, healthy
imply, infer
in, into
linear, lineal
loose, lose
moral, morale
mutual, common
passed, past
percent, percentage
practical, practicable
precede, proceed
predominant, predominate
principal, principle
rare, scarce, unique
sewage, sewerage
suspense, suspension
there, their, they're
unsolvable, insoluble
valuable, valued

Exercises

A. *Be prepared to define and illustrate the proper meanings of the words listed above. Refer to your dictionary if necessary.*

B. *Detect the misused words in the following sentences and make the corrections needed.*

1. The alumnae of Wabash College are all men.
2. Mrs. Malaprop said, "I would by no means wish a daughter of mine to be a progeny of learning."
3. The mayor, the principle man of the town, was also president of the bank.

4. The usage of these numbers is very inconsistent in that both the figure and the written numeral are used at one time or another.
5. The base of the ring stand is commonly 8.0 inches x 6.0 inches in length and breadth respectively.
6. The continuous use of a word in every sentence is monotonous.
7. Some new houses affront the opposite side of the street.
8. The commander had less men than in his previous campaign.
9. Of all the spiders the one that lives in a bubble under the water is the most unique.
10. The well produced a continual flow of oil.
11. His remarks inferred that he did not believe the report.
12. All editing chores were divided between the three editors.
13. The materials used in the construction of the building and the layout of the apartment are included in the report.
14. Because of the amount of people smoking in the room the air was full of smoke.
15. What would be the affect of doubling the amount of sewerage emptied in the river?
16. The problem that you pose concerning the balance of your data seems to be insoluble.
17. The Wright Brothers discovered the first practical airplane in 1905.
18. The morale of the men working in the mines came to be so bad that the per cent of absences was intolerant.
19. The predominate feeling among the workers was that management had no sympathy whatever for their complaints.
20. The middle of the circle is the point exactly equidistant from all other points in the circumference.

ILLOGICAL SENTENCE STRUCTURE

Carelessly constructed sentences sometimes convey a meaning different from that intended. Faulty comparisons like the following are frequent:

1. The wire of which the circuit is made is larger in diameter than the ordinary light bulb.

Here the comparison is incompletely expressed: a wire of greater diameter than a light bulb would be large indeed. The sentence should read:

The wire of which the circuit is made is larger than that of the ordinary light bulb.

Since a wire could not be larger except "in diameter," that phrase is superfluous.

2. The projector for the many "stars" is a piece of work comparable in its own field to that of the 200-inch telescope.

Since the antecedent of "that" is "a piece of work," the writer says that the projector is comparable to the piece of work of the telescope. But the projector is a device, not a piece of work.

The projector for the many "stars" is an instrument comparable in its own field to the 200-inch telescope.

3. Your description of the trouble you have had is not serious and can be cured by some minor adjustments.

Here the grammatical subject is "description," but the logical subject is "trouble."

> From your description I conclude that your trouble is not serious and can be avoided if you make some minor adjustments.

Overuse of Technical Terms

When used with skill and restraint, technical terms make for clearness and conciseness. When used carelessly and too freely, however, they result in such writing as *The New Yorker* exhibited for the amusement of its reader in the following passage:

> In a machine of the character described, the combination of an array of totalizer wheels arranged to be propelled through multiple predetermined circumferential superpositions; means for variously measuring the magnitude of optionally determinable superpositions composing a pre-arranged taxonomy of a consecutive succession of denominational manipulative devices of substantially minimum dactylonomy; an intrinsically heterogeneous precomputed taxonomy of abutments controlled thereby arranged in nonconsecutive monodromic sequence and in denominational seriatim concatenation; and means cooperating with the aforesaid abutments to transvert the heterogeneous taxonomy into a compensated sequence of multiple equiangular predesigned magniudes compatible with said circumferential superpositions substantially as described.—Coxhead & Norton Patent, U. S. Patent No. 1986137, Original Claim No. 22.

Careful writers limit their use of such language even when they are sure that their readers will understand it. When your readers may be unfamiliar with technical terms, either define them or substitute nontechnical words. To use technical terms merely to impress readers is inexcusable—especially when the sentence is grammatically faulty like the one quoted.

The following definition of Portland cement, Designation C 9-30 of the American Society for Testing Materials, justifiably includes technical terms because it is intended for engineers familiar with them:

> Portland cement is the product obtained by finely pulverizing clinker produced by calcining to incipient fusion an intimate and properly proportioned mixture of argillaceoús and calcareous materials, with no additions subsequent to calcination except water and calcined or uncalcined gypsum.

The definition in *Webster's New International Dictionary*, Second Edition, intended for the general reader, is less technical:

> A widely used hydraulic cement made by burning and grinding a carefully proportioned artificial mixture of pure limestone and clay, or of other aluminous material, or of aluminous limestone and pure limestone—so called from the resemblance of concrete made with it to Portland stone. . . .—By permission. From *Webster's New International Dictionary, Second Edition*, © 1961 by G. & C. Merriam Co., Publishers of the Merriam-Webster Dictionaries.

The definition from the Third Edition (1961) of the same dictionary is much shorter and more technical:

> A hydraulic cement made by finely pulverizing the clinker produced by calcining to incipient fusion a mixture of argillaceous and calcareous materials.—By permission. From *Webster's Third New International Dictionary*, © 1976 by G .& C. Merriam Co., Publishers of the Merriam-Webster Dictionaries.

One further illustration of the excessive use of technical terms is quoted from the G. & C. Merriam Company publication *Word Study* for May, 1946, which in turn quoted it from the Camp Livingston *Communiqué*:

> Someone has wired a government bureau asking whether hydrochloric acid could be used to clean a given type of boiler tube. The answer was: "Uncertainties of reactive process make use of hydrochloric acid undesirable where alkalinity is involved." The inquirer wrote back, thanking the bureau for the advice and adding that he guessed he would use hydrochloric acid. The bureau wired him: "Regrettable decision involves uncertainties. Hydrochloric acid will produce submuriate invalidating reactions." Again the man wrote, thanking them for their advice and adding that he was glad to know that hydrochloric acid was all right. This time the bureau wired in plain English: "Hydrochloric acid," said the telegram, "will eat h . . . out of your tubes."

The waste here is self-evident.

Exercises

A. *In one of your textbooks find passages that seem to you unnecessarily technical and bring them to class for analysis and criticism.*

B. *Revise the following sentences so that the intended meanings are clear:*

1. It is a common known fact, however, that the growth rate of spruce and tamarack in bog areas is slower than aspen on poor areas.
2. There is some evidence that when the body is in an edematous state, that sodium ions enter the cell.
3. There was no attempt made to determine why this was a problem, but only with elimination of future difficulties with this problem.
4. Each road system was compiled, subtotaled, and totaled by column.
5. It seems important to realize that just because one has found a red soil in the tropics that it belongs to this great soil group.
6. Each piece of data will be compared to the existing speed, weather, and road conditions.
7. An adequate description of a sedimentary rock may be obtained by four characteristics, color, texture, structure, and fossils.
8. Voltage is similar to the pump, and current is similar to the water.
9. The fact that the air will take the path of least resistance is, in this case, spreading undesirable odors.
10. Because the most common method of arranging a large number of words is the alphabetical arrangement, the common dictionary is so constructed.
11. The relative success or failure of these investments is based on the decision of financial analysis of the financial statements.

12. The around-the-bend baffles, as they are called, are actually the channels bent around to flow in the opposite direction.
13. In 1961 Dr. Sauberlich reported a study in which he added amino acid mixture to rats on a protein-free diet.
14. The fact that threonine is most effective in reducing liver fat deposition in rats fed certain proteins than it is when others are fed, even when the actual amounts of threonine are the same, indicates that other amino acids are involved.
15. The blueprints and materials used to build the pens are shown in Appendix C.
16. The age of an ice area is the best way to determine whether the ice is safe for skating.
17. Rafters should be placed 16 inches between centers but can be varied somewhat by using larger members.
18. Would you submit to us the location, layout and design for our new plant.
19. The differences of the Quakers' way of life certainly varies from the majority of people in the United States.
20. The nature of edema caused by malnutrition is different from edema carried by disease.

MISCELLANEOUS FAULTS

In each of the preceding sections in this chapter, all sentences in the exercise have had the same fault, though many had other faults as well. When sentences are grouped in this way, the faults are fairly easy to identify.

When faults are unclassified, however, a more discriminating analysis is required. This section therefore is added in which the exercise is made up of sentences having a variety of faults.

Exercise

The sentences below have faults of various kinds. Write a brief criticism of each sentence and then revise the sentence.

1. The panel is usually selected at random, frequently does not contain trained evaluators, and often is held in poor evaluating conditions.
2. One group of calves would be on strictly a milk (M) diet and the other main group would be on a normal ration of milk, hay, and grain (MHG).
3. A perusal of the literature reveals very little research conducted in regards to atherosclerosis and the relationship of feeding cholesterol to bovine and subsequent atherogenesis.
4. My company has used similar approaches to similar problems for other companies and they have proved successful.
5. As everywhere else, Pennsylvania will be holding elections soon.
6. The reason for compilation of this section's data first was due to the completeness of data provided by this section.

7. There appears to be no correct answer for the situation, but remains a searching process for the purpose of developing an equitable method of disclosure.
8. If your growth is typical of many other firms, your organization took place in the manner of a jigsaw puzzle.
9. In addition to structural problems, foundation studies must be made to insure that the tower will not settle or tip.

Notes

1. Grammatical terms are used occasionally in this chapter without definitions. If you do not know the meaning of a term, look it up in your dictionary or a freshman English handbook.

Bibliographies

In no instances are any of the following lists exhaustive. They are here to provide students with places to begin looking and to suggest some of the riches available in any large library. For example, approximately 60 items are listed here under "Abstracting Journals and Indexes," but a typical major university's library may actually carry more than 600.

THE METHOD OF SCIENCE

If you would like more information about science, its methods, history, and vocabulary, the following will be helpful :

Achinstein, Peter, *Concepts of Science: A Philosophical Analysis,* Baltimore: Johns Hopkins University, 1968.
Adamson, R., and H. A. Jevons (eds.), *Pure Logic and Other Minor Works,* New York: B. Franklin Press, 1971.
Beardsley, Monroe C., *Thinking Straight,* New York: Prentice-Hall, Inc., 1975.
Beveridge, W.I.B., *The Art of Scientific Investigation,* London: W. Heineman, 1957.
Blake, Ralph M., Curt J. Ducasse and Edward H. Madden, *Theories of Scientific Method: The Renaissance Through The Nineteenth Century,* Seattle: University of Washington Press, 1970.
Cohen, Morris R., and Earnest Nagel, *An Introduction to Logic and Scientific Method,* New York: Harcourt Brace Jovanovich, Inc., 1962.
Conant, James B., *Science and Common Sense,* New Haven: Yale University Press, 1951.
Copi, Irving M., *Introduction to Logic,* New York: The Macmillan Company, 1978.
Davis, J. T., *The Scientific Approach,* New York: Academic Press, 1965.
Kuhn, Thomas S., *The Structure of Scientific Revolutions,* Second Edition Enlarged, Chicago: University of Chicago Press, 1970.
Langride, Derek (ed.), *The Universe of Knowledge,* College Park: University of Maryland, 1969.
Levi, Isaac, *Gambling with Truth,* New York: Alfred E. Knopf, 1974.
Mumford, Lewis, "Reflections (The Twentieth Century)," *The New Yorker,* March 10, 1975, pp. 42 ff.
Nidditch, Peter H., *Introductory Formal Logic of Mathematics,* Boston: Routledge and Kegan Paul, 1971.
Pearson, K., *The Grammar of Science,* New York: Meridian Books, Inc., 1957.
Rapport, Samuel B. (ed.), *Science: Method and Meaning,* New York: New York University Press, 1963.

THE PROCESS OF WRITING

The following books describe various stages of the writing process:

Corder, Jim W., *Contemporary Writing,* Glenview, Illinois: Scott, Foresman, 1979, Unit I, "The Process of Writing."
Cowan, Gregory, and Elizabeth Cowan, *Writing,* New York: John Wiley & Sons, 1980.
D'Angelo, Frank J., *Process and Thought in Composition,* Second Edition, Cambridge, Mass.: Winthrop, 1980.
Hairston, Maxine, *Successful Writing,* New York: Norton, 1981.
Harty, Kevin J., *Strategies for Business and Technical Writing,* New York: Harcourt Brace Jovanovich, 1980.

Lanham, Richard, *Revising Prose,* New York: Charles Scribner's Sons, 1979.
Murray, Donald M., *A Writer Teaches Writing,* Boston: Houghton Mifflin, 1968.
Trimble, John R., *Writing with Style,* Englewood Cliffs, New Jersey: Prentice-Hall, 1975.
Weil, Benjamin H., *Technical Editing,* Westport, Connecticut: Greenwood Press, 1975.

AUDIENCE ANALYSIS AND ADAPTATION

Anderson, Paul V., ed., *Teaching Technical Writing: Teaching Audience Analysis and Adaptation,* Anthology No. 1, n.p.: The Association of Teachers of Technical Writing, 1980.
Boettinger, Henry M., *Moving Mountains,* New York: Macmillan, 1969.
Cowan, Gregory, and Elizabeth Cowan, *Writing,* New York: John Wiley & Sons, 1980, pp. 80-93.
Mathes, J. C., and Dwight W. Stevenson, *Designing Technical Reports,* Indianapolis: Bobbs-Merrill, 1976.
Morrisey, George L., *Effective Business and Technical Presentations,* Second Edition, Reading, Mass.: Addison-Wesley, 1975.
Pearsall, Thomas E., *Audience Analysis for Technical Writing,* Beverly Hills: Glencoe Press, 1969.
Simons, Herbert W., *Persuasion: Understanding, Practice, and Analysis,* Reading, Mass.: Addison-Wesley, 1976.

VISUALS

Bowman, William J., *Graphic Communications,* New York: John Wiley and Sons, Inc., 1968.
Blicq, R. S., *Technically Write,* Ch. 9, "Illustrating Technical Documents," Englewood Cliffs, N. J.: Prentice-Hall, Inc., 1972.
Fear, David E., *Technical Writing,* Ch. 6, "Illustrations," New York: Random House, Inc., 1978.
Haemer, K. W., *Making the Most of Charts: An ABC of Graphic Presentation,* New York: American Telephone and Telegraph, 1960.
Hanks, Kurt, and Larry Belliston, *Draw,* Los Altos, Cal.: W. Kaufmann, 1977.
Houp, Kenneth W., and Thomas Pearsall, *Reporting Technical Information,* Part 2, Ch. 12, "Graphical Elements," Beverly Hills: Glencoe Press, 1977.
Levens, A. S., *Graphics.* New York: John Wiley and Sons, Inc., 1968.
Magnan, George A., *Using Technical Art,* New York: John Wiley and Sons, Inc., 1970.
Sklare, Arnold B., *The Technician Writes: A Guide to Basic Technical Writing,* Ch. 10, "Visual Aids," San Francisco: Boyd and Fraser Publishing Company, 1971.
Strong, Charles W., and Donald Edison, *A Technical Writers Handbook,* Ch. 12, "Displaying Data: Preparation of Graphic Aids," Ch. 13, "Displaying Data: Types of Graphic Aids," New York: Holt, Rinehart and Winston, Inc., 1971.
Ulman, Joseph N., and Jay R. Gould, *Technical Reporting,* Section III, Ch. 19, "Visual Presentation of Information," New York: Holt, Rinehart and Winston, Inc., 1972.
Weisman, Herman M., *Basic Technical Writing,* Ch. 7, "Graphic Presentation in Technical Writing," Columbus: Charles E. Merrill Publishing Company, 1974.

CLASSIFICATIONS

For the student who wants more information concerning classification, the following will be helpful:

Beardsley, Monroe C., *Practical Logic,* Part II, Ch. 12, "Sorting and Sampling," New York: Prentice-Hall, Inc., 1950.

Beardsley, Monroe C., *Thinking Straight,* Ch. 3, Section 9, "Classification," New York: Prentice-Hall, Inc., 1975.

Cohen, Morris R., and Earnest Nagel, *An Introduction to Logic and Scientific Method,* Ch. XII, "Classification and Definition," New York: Harcourt, Brace and Company, 1962.

Coleman, Peter, and Ken Brambleby, *The Technologist as Writer,* "Definitions and Classifications," New York: McGraw Hill Book Co., 1971.

Hayakawa, S. I., *Language in Thought and Action,* Ch. 12, "Classification," New York: Harcourt, Brace and Company, 1978.

Houp, Kenneth W., and Thomas F. Pearsall, *Reporting Technical Information,* "Classification and Division," pp. 86-90, Beverly Hills: Glencoe Press, 1977.

Langridge, Derek (ed.), *The Universe of Knowledge,* College Park: University of Maryland, 1969.

Laster, Ann A., and Nell Ann Pickett, *Writing for Occupational Education,* Ch. 4, "Analysis Through Classification and Partition: Putting Things in Order," San Francisco: Canfield Press, 1974.

Strong, Charles W., and Donald Edison, *A Technical Writer's Handbook,* Ch. 3, "Techniques of Technical Writing: Classification," New York: Holt, Rinehart and Winston, 1971.

Weisman, Herman M., *Basic Technical Writing,* "Classification," pp. 257-261, Columbus: Charles E. Merrill Publishing Company, 1974.

ANALYSES

For a more detailed understanding of analysis the following will be helpful:

Brand, Norman, and John O. White, *Legal Writing,* New York: St. Martin's Press, 1976, "Linear Analysis."

Brown, Leland, *Effective Business Report Writing,* Part II, "Steps in the Preparation of Business Reports," New York: Prentice-Hall, 1973.

Coleman, Peter, and Ken Brambleby, *The Technologist as Writer,* Ch. 2, "Tests of Evidence," New York: McGraw Hill Book Company, 1971.

Johnson, James William, *Logic and Rhetoric,* Ch. VII, "Logical Thinking," New York: The Macmillan Company, 1962.

Laster, Ann A., and Nell Ann Pickett, *Writing for Occupational Education,* Ch. 5, "Analysis Through Effect and Cause: Answering Why," San Francisco: Canfield Press, 1974.

Martin, Harold C., and Richard M. Ohmann, *The Logic and Rhetoric of Exposition,* Part One, "Proving," New York: Holt, Rinehart and Winston, Inc., 1963.

Pauley, Steven, *Technical Report Writing Today,* Ch. 5, "Interpreting Statistics," Boston: Houghton Mifflin Company, 1973.

Ruggiero, Vincent Ryan. *Beyond Feelings: A Guide to Critical Thinking,* Ch. 20, "Interpreting Evidence," Ch. 21, "Analyzing A Position," New York: Alfred Publishing Company, Inc., 1975.

Strong, Charles W., and Donald Eidson, *A Technical Writer's Handbook,* Ch. 10, "Evaluating Data: Logic," Ch. 11, "Evaluating Data: Statistics," New York: Holt Rinehart and Winston, Inc., 1971.
Toulmin, Stephen, and Richard Rieke, and Allan Janik, *An Introduction to Reasoning,* New York: Macmillan, 1979.
Weisman, Herman M., *Basic Technical Writing,* Ch. 10, "Special Expository Techniques in Technical Writing–Description, Explanation and Analysis," Columbus: Charles E. Merrill Publishing Company, 1974.
Wicker, C. V., and W. P. Allbrecht, *The American Technical Writer,* Ch. 7, "Analysis," Ch. 8, "Analysis of Mechanisms and Processes," New York: The American Book Company, 1960.

REQUISITES IN REPORT WRITING

Blicq, R. S., *Technically Write,* Englewood Cliffs, N. J.: Prentice-Hall, Inc., 1972.
"Presents both formal and informal writing problems by confronting students with various positions in two technically oriented industries."
Brogan, John A., *Clear Technical Writing,* New York: McGraw-Hill Book Company, 1973.
"A programmed learning text which gives common problems of technical writing and exercises to correct them."
Coleman, Peter, and Ken Brambleby, *The Technologist As Writer,* New York: McGraw-Hill Book Company, 1971.
"Presents report writing fundamentals and case studies ranging from simple requests to formal reports."
Ehrlich, Eugene, and Daniel Murphy, *The Art of Technical Writing* (*A Manual for Scientists, Engineers, and Students*), New York: Thomas Y. Crowell Company, 1969.
"A brief treatment of types, with strong emphasis on style and usage."
Fear, David E., *Technical Writing,* New York: Random House, 1978.
"Aimed at students with little interest in English composition, it offers models representing writing done in practical work situations."
Glidden, H. K., *Reports, Technical Writing, and Specifications,* New York: McGraw-Hill Book Company, 1964.
"Covers the entire range of technical writing; contains a little bit of everything."
Graves, Harold F., and Lyne S. S. Hoffman, *Report Writing,* Englewood Cliffs, N.J.: Prentice-Hall, Inc., 1965.
"Details of the techniques applicable to every report-writing situation. Six specimen reports are reproduced."
Hays, Robert, *Principles of Technical Writing,* Reading, Mass.: Addison-Wesley Publishing Company, Inc., 1965.
"Concentrates on the beginning writer with attention on clarity of style."
Hicks, Tyler G., *Writing for Engineering and Science,* New York: McGraw-Hill Book Company, 1961.
"Covers both reports and other types of technical writing; contains both printed and typed examples."
Houp, Kenneth W., and Thomas E. Pearsall, *Reporting Technical Information,* Beverly Hills: Glencoe Press, 1977.
"Gives thorough explanation of basic principles and applications in easily understood terms."

Laster, Ann A., and Nell Ann Pickett, *Writing for Occupational Education,* San Francisco: Canfield Press, 1974.

"Emphasizes utilitarian practical aspects simply. Oriented to technician in two-year program, but helpful for any level."

Lesikar, Raymond V., *Report Writing for Business,* Homewood, Ill.: Richard D. Irwin, Inc., 1977.

"Gives major emphasis to organizing and writing reports and stresses research methodology."

Mathes, J. C., and Dwight W. Stevenson, *Designing Technical Reports,* Indianapolis: Bobbs-Merrill, 1976.

"Grounded on the axiom that 'the needs of the audiences in the organizational system determine the design of the report.' "

Mills, Gordon H., and John A. Walter, *Technical Writing,* New York: Holt, Rinehart and Winston, 1978.

"A thorough treatment of the style, forms, and organization of reports, with much illustrative material."

Pauley, Steven E., *Technical Report Writing Today,* Boston: Houghton Mifflin Company, 1973.

"Emphasizes overcoming communication gap between specialist and non-specialist and explains various techniques."

Ross, Peter Burton, *Basic Technical Writing,* New York: Thomas Y. Crowell Company, 1974.

"Designed to acquaint student with methods of analyzing, organizing, and presenting technical material. Uses simplistic 'Dick and Jane' approach to clarify, exemplify material."

Schultz, Howard, and Robert C. Webster, *Technical Report Writing,* New York: David McKay Company, Inc., 1962.

"Describes fifty problems typical of those in reports and presents solutions."

Tallent, Norman, *Psychological Report Writing,* Englewood Cliffs, N. J., 1976.

"A text on problems of psychological reports and their preparation."

Tichy, H. J., *Effective Writing for Engineers, Managers, Scientists,* New York: John Wiley and Sons, Inc., 1967.

"Good instruction on matters of style, diction, and sentence structure, telling how to write more effectively and easily."

Turner, Rufus P., *Technical Report Writing,* San Francisco: Rinehart Press, 1971.

"The book provides rudiments of technical report writing and standard professional practices along with illustrative examples and exercises."

Van Hagen, Charles E., *Report Writer's Handbook,* Englewood Cliffs, N. J.: Prentice-Hall, Inc., 1961.

"A practical guide to effective techniques in writing technical reports. Intended for the professional."

Ward, Ritchie R., *Practical Technical Writing,* New York: Alfred A. Knopf, 1968.

"Presents the results of a survey to ascertain what managers expect in a good technical report."

Weisman, Herman M., *Basic Technical Writing,* Columbus, Ohio: Charles E. Merrill Publishing Company, 1974.

"Describes a comprehensive and unified approach to the problems of technical writing; contains two course outlines useful for beginning instructors."

BUSINESS LETTERS

Damerst, William A., *Resourceful Business Communication,* New York: Harcourt, Brace and World, Inc., 1966.

Devlin, Frank J., *Business Communication,* Homewood, Ill.: Richard D. Irwin, Inc., 1968.

Himstreet, William C., and Wayne M. Baty, *Business Communications: Principles and Methods,* San Francisco: Wadsworth Publishing Company, 1977.

Janis, J. Harold, *Writing and Communication in Business,* New York: The Macmillan Company, 1978.

Lesikar, Raymond V., *Business Communication: Theory and Application,* Homewood, Illinois: Richard D. Irwin, Inc., 1976.

Parkhurst, Charles C., *Business Communication for Better Human Relations,* Englewood Cliffs, N.J.: Prentice-Hall, Inc., 1966.

Poe, Roy W., and Rosemary T. Fruehling, *Business Communications: A Problem Solving Approach,* New York: McGraw-Hill Book Company, 1978.

Sigband, Norman, *Communication for Management and Business,* Glenview, Ill.: Scott, Foresman and Company, 1976.

Weeks, Francis W., and Daphne A. Jameson, *Principles of Business Communication,* Champaign, Ill.: Stipes Publishing Company, 1979.

Wilkinson, C. W., Peter B. Clarke, and Dorothy C. M. Wilkinson, *Communicating Through Letters and Reports,* Seventh Edition, Homewood, Illinois: Richard D. Irwin, Inc., 1980.

PROPOSALS

Ehrlich, Eugene, and Daniel Murphy, *The Art of Technical Writing,* Ch. 2, "The Technical Proposal," New York: Thomas Y. Crowell Company, 1969.

Hill, William J., *Successful Grantsmanship,* Steamboat Springs, Colorado: Grant Development Institute, 1977.

Holtz, Herman, *Government Contracts: Proposalmanship and Winning Strategies,* New York: Plenum Press, 1979.

Houp, Kenneth W., and Thomas E. Pearsall, *Reporting Technical Information,* Part 3, Ch. 15, "Proposals," Beverly Hills: Glencoe Press, 1977.

Mandel, Siegfried, *Writing for Science and Technology,* Ch. 13, "Proposed Writing," New York: Dell Publishing Company, 1970.

Mills, Gordon H., and John A. Walter, *Technical Writing,* Ch. 15, "Proposals," Holt, Rinehart and Winston, Inc., 1978.

Pauley, Steven E., *Technical Report Writing Today,* Ch. 10, "Proposals," Boston: Houghton Mifflin Company, 1973.

Proposals . . . and their Preparation, Washington, D.C.: Society for Technical Communication, 1974.

Ross, Peter Burton, *Basic Technical Writing,* Ch. 8, "The Proposal," New York: Thomas Y. Crowell Company, 1974.

Ulman, Joseph N., and Jay R. Gould, *Technical Reporting,* Section II, Ch. 11, "Proposals," Holt, Rinehart and Winston, Inc., 1972.

LIBRARY RESOURCES

In addition to the *subject* section of the library card catalogue, there are innumerable other sources of information. A few of those which the student is likely to find most useful are listed below:

Encyclopedias and Dictionaries

Britannica Yearbook of Science and the Future
Dictionary of the Social Sciences
Encyclopedia Americana
Encyclopaedia Britannica
Encyclopedia of the Biological Sciences
Encyclopedic Dictionary of Physics
Encyclopedia of Chemical Technology
Encyclopedia of Science and Technology

Reference Books

Facts on File
Guide to Reference Books
Guide to the Use of Books and Libraries
Guide to the World's Abstracting and Indexing Services in Science and Technology
World Almanac

Guides to Books and Book Reviews

Books in Print (BIP) (1948-)
Book Review Digest (1905-)
Cumulative Book Index (CBI) (1928-)
Technical Book Review Index (1935-)
U.S. Catalog of Books
U.S. Quarterly Book List (1945-)

Guides to Periodical Literature

Business Periodicals Index
Industrial Arts Index (1913-1958)
International Index to Periodicals
Readers' Guide to Periodical Literature
Ulrich's International Periodicals Directory

Abstracting Journals and Indexes

Abstracts of Health Effects of Environment
Abstracts of North American Geology
Accountants' Index
Aerospace Engineering Review
Agricultural and Horticultural Abstracts
Agricultural Index

Animal Breeding Abstracts
Applied Ecology Abstracts
Applied Science and Technology Index
Art Index
Astronomy and Astrophysics Abstracts
Biological Abstracts
Biological and Agricultural Index
British Technology Index
Business Periodicals Index
Ceramic Abstracts
Chemical Abstracts
Computer and Control Abstracts
Computer and Information Systems
Current Contents (for various fields)
Dairy Science Abstracts
Dissertation Abstracts International
Economic Abstracts
Education Index
Electrical and Electronics Abstracts
Electronics and Communications Abstracts
Engineering Abstracts
Engineering Index
Environment Abstracts and Index
Excerpta Medica
Forestry Abstracts
Geoscience (formerly *Geological*) *Abstracts*
Highway Research Abstracts
Index Medicus
Information Science Abstracts
International Abstracts of Biological Sciences
International Aerospace Abstracts
Mathematical Review
Metallurgical Abstracts
Metals Abstracts
Meteorological and Geoastrophysical Abstracts
Mineralogical Abstracts
Nuclear Science Abstracts
Nutrition Abstracts and Reviews
Petroleum Abstracts
Physics Abstracts
Plant Breeding Abstracts
Pollution Abstracts
Psychological Abstracts
Public Affairs Information Service
Reviews of Modern Physics
Science Abstracts
Science Citation Index
Scientific and Technical Aerospace Reports (*STAR*)
Sociological Abstracts
Statistical Abstract of the United States
U.S. Government Research Reports
Wildlife Review
Zoological Record

Computerized Data Bases and Bibliographies

ABI/INFORM—Abstracted Business Information, 1971-
ABM—Art Bibliographies Modern
ACCOUNTANT'S INDEX—Accounting literature file from AICPA, 1974-
AGRICOLA—World agricultural literature from National Agricultural Library, 1970-
AHL—America: History and Life
AIM/ARM—Vocational-Technical Education Materials, 1967-1976
API—American Petroleum Institute's literature and patent file, 1964-
APTIC—Air Pollution Abstracts, 1966-
ASFA—Aquatic Sciences and Fisheries Abstracts, 1975-
ASI—American Statistics Index (U. S. government statistical publications), 1974-
AVLINE—Reviews of medical educational media, 1975-
BIOSIS—Biological Abstracts Previews, 1969-
CAB—Commonwealth Agricultural Bureaux, 1972-
CANCERLINE—Cancer literature
CBPI—Canadian Business Periodicals Index, 1975-
CDI—Comprehensive Dissertation Index, 1861-
CHEMCON—Chemical Abstracts Condensates (supplemented by CHEM 70/71), 1970-
CHEMDEX, CHEMLINE, CHEMNAME } online chemical dictionaries, 1972-
CIN—Chemical Industry Notes from CAS, 1974-
CIS—Congressional Information Service (Index to congressional documents), 1970-
CLAIMS—U.S. chemical and chemically related patents, 1950-
CNI—Canadian News Index, 1977-
COMPENDEX—Engineering Index, 1970-
CONFERENCE PAPERS INDEX—Papers & proceedings from professional meetings, 1973-
CRECORD—Congressional Record, 1976-
CRIS—Current Research Information System of U.S.D.A., 1975-
DISCLOSURE—Index to official corporate SEC filings, 1978-
DMMS—Defense Market Measures System, 1975-
EC ERIC—Exceptional Children Abstracts, 1969-
EDB—Energy Data Base of E.R.D.A.
EIS—Directory/sales information about U.S. firms
ENERGYLINE—Energy Information Abstracts, 1971-
ENVIROLINE—Environment Abstracts, 1971-
EPB—Environmental Periodicals Bibliography, 1973-
EPILEPSYLINE—Epilepsy literature
ERIC—Educational Resources Information Center, 1966-
EXCERPTA MEDICA—Biomedical information, 1975-
F&S—Funk & Scott Index of Industries and Corporations, 1972-
FEDERAL INDEX—Citations from Fed. Register, Congressional Record, etc., 1977-
FED REG—Federal Register, 1977-
FOUNDATIONS—Foundations Directory/Foundations Grants Index, 1973-
FSTA—Food Science and Technology Abstracts, 1972-
GEOARCHIVES—Geological literature
GEOREF—American Geological Institute's Reference File, 1967-

GRANTS—Current governmental and private sector grants
HA—Historical Abstracts, 1973-
INSPEC—Science Abstracts (Physics, Electronics, computers), 1969-
IPA—International Pharmaceutical Index, 1970-
ISMEC—Mechanical Engineering/Engineering Management file of IEE & Inst. Mech. Eng., 1973-
LABORDOC—Labor/industrial relations literature in the ILO files, 1965-
LIBCON—Library of Congress cataloging information, 1965-
LISA—Library & Information Science Abstracts, 1969-
LLBA—Language and Language Behavior Abstracts, 1973-
MAGINDEX—Magazine Index for general periodicals, 1977-
MANAGEMENT CONTENTS—Article citations from business/management journals, 1974-
MEDLINE—National Library of Medicine's MEDLARS system on-line, 1966-
METADEX—Metals Abstracts/Alloys Index, 1966-
MGA—Meterlogical and Geoastrophysical Abstracts, 1972-
MLA—International Bibliography from the Modern Language Assn., 1976-
MONTHLY CATALOG—Catalog of U.S. government publications, 1976-
MOS—Energy and Environmental Models Survey Data Base
MRIS—Abstracts from Marine Research Information Service, 1970-
NICEM—National Information Center for Educational Media, 1977-
NICSEM—Media and devices for handicapped, 1974-
NRC—National Referral Center for information centers
NSA—Nuclear Science Abstracts, 1967-1976
NSC—Nuclear Safety Information
NSR—Nuclear Structure Data Base
NTIS—National Technical Information Service (government sponsored research reports), 1964-
OCEANIC ABSTRACTS—Oceanography and marine related materials, 1964-
P/E NEWS—Petroleum industry news file of American Institute, 1975-
PAIS—Public Affairs Information Service, 1976- (English); 1972- (foreign)
PAPERCHEM—Institute of Paper Chemistry, 1968-
PIRA—Paper; Printing, Packaging literature, 1975-
POLLUTION—Pollution Abstracts, 1970-
PREDICASTS—Statistical services dealing with wide range of U.S. business concerns, 1971-
PROMT—Chemical and Equipment Market Abstracts, 1972-
PSYCHABS—Psychological Abstracts, 1967-
QUEBEC—French language news index, 1978-
SAE—Society of Automotive Engineers Abstracts, 1965-
SAFETY—Safety Abstracts, 1978-
SCISEARCH—Science Citation Index, 1974-
SOCIAL SCISEARCH—Social Science Citation Index, 1972-
SA—Sociological Abstracts, 1963-
SPIN—Physics literature notices,, 1975-
SSIE—Smithsonian Science Information Exchange (current research projects)
TITUS—Technical textile literature, 1970-
TULSA—Petroleum Abstracts, 1965-
USPSD—Political Science literature, 1975-
WAA—World Aluminum Abstracts, 1968-
WRA—Water Resources Abstracts of U.S. Dept. of Interior
WTA—World Textile Abstracts, 1970-

Guides to Writing Research Papers

For more information about the writing of papers requiring library research, consult one of the following:

Barton, Mary N. *Reference Books, A Brief Guide for Students and Other Users of the Library,* Baltimore: Enoch Pratt Free Library, 1970.
Coyle, William, *Research Papers,* New York: The Odyssey Press, 1976.
Hook, Lucyle, and Mary Virginia Gaver, *The Research Paper,* Englewood Cliffs, N.J.: Prentice-Hall, Inc., 1969.
Hutchinson, Helene D., *The Hutchinson Guide to Writing Research Papers,* Beverly Hills: Glencoe Press, 1973.
Lester, James D., *Writing Research Papers: A Complete Guide,* Glenview, Ill.: Scott, Foresman and Company, 1980.
Rubenstein, S. Leonard, et al., *Writing The Research Paper,* Boston: Allyn and Bacon, Inc., 1969.
Sears, Donald A., *Harbrace Guide to the Library and Research Paper,* New York: Harcourt, Brace and Company, 1973.
Shores, Louis, *Basic Reference Sources,* Chicago: American Library Association, 1972.

MECHANICAL ASPECTS OF STYLE

For students who wish additional information on style and usage, here are some titles from which to choose:

Selected Dictionaries for Student Use

American Heritage Dictionary, New York: Houghton Mifflin, 1978.
The most recent, comprehensive, and readable of the desk dictionaries. Its usage notes are especially helpful.
Chamber's Technical Dictionary, C. F. Tweney (ed.), New York: The Macmillan Company, 1958.
Fully supplemented, useful for both specialist and student.
Compton's Dictionary of Technical Terms (two volumes), New York: Compton, 1966.
Simply written for nonspecialist.
The Basic Dictionary of Science, E. C. Graham (ed.), New York: The Macmillan Company, 1966.
Over 25,000 words basically defined.
Webster's Third New International Dictionary, Springfield, Mass.: G. & C. Merriam Company, 1976.
The most authoritative and complete American dictionary.

Selected Material on Style and Usage

A Manual of Style, 12th ed., Chicago: University of Chicago Press, 1969.
A standard authority and invaluable reference for all persons concerned with getting words into print.
Adams, Dorothy, and Margaret Kurtz, *The Technical Secretary: Terminology and Transcription,* New York: McGraw-Hill Book Company, 1968.

Concerned primarily with technical shorthand, it also gives stylistic practices used in typing formulae, equations, abbreviations and terms in technical reports.

Bernstein, Theodore M., *The Careful Writer, A Modern Guide to English Usage,* New York: Atheneum Publishers, 1965.

More than 2,000 alphabetized entries . . . on questions of use, meaning, grammar, punctuation, precision, logical structure, and color.

Brusaw, Charles T., Gerald J. Alred, and Walter E. Oliu, *Handbook of Technical Writing,* New York: St. Martin's Press, 1976.

A comprehensive reference guide for students in technical writing courses.

Brusaw, Charles T., Gerald J. Alred, and Walter E. Oliu, *The Business Writer's Handbook,* New York: St. Martin's Press,

A comprehensive reference guide for business writing.

Carey, G. V., *Mind the Stop,* New York: Penguin Books, 1971.

A brief guide to punctuation, descriptive of British usage.

Dodds, Robert H., *Writing for Technical and Business Magazines,* New York: John Wiley & Sons, 1969.

A step-by-step discussion of writing and placing articles.

Evans, Bergen, and Cornelia Evans, *A Dictionary of Contemporary American Usage,* New York: Random House, 1967.

A ready-reference guide to the effective use of the English language, listing word preferences, grammar, style, punctuation, idiom, etc., based on modern linguistic scholarship.

Fishbein, Morris, *Medical Writing,* Chicago: American Medical Association, 1978.

Dependable guide for preparation of material for the *Journal of the American Medical Association.*

Follett, Wilson, *Modern American Usage,* New York: Hill and Wang, Inc., 1974.

An American book of usage grounded in the philosophy that the best in language —which is often the simplest—is not too good to be aspired to.

Fowler, H. W., and F. G. Fowler, *The Kings English,* Fair Lawn, N.J.: Oxford University Press, 1974.

An old standby in answering questions of usage and style. Illustrates common problems by example.

Hutchinson, Lois Irene, *Standard Handbook for Secretaries,* 8th ed., New York: McGraw-Hill Book Company, 1971.

The eighth edition of a widely used guide to correlate business practice with printing practice.

Mitchell, John H., *Writing for Professional and Technical Journals,* New York: John Wiley & Sons, 1968.

Defines the characteristics of technical articles and anthologizes conflicting sections of style guides issued for various disciplines.

Monroe, Kate, et al., *The Secretary's Handbook,* New York: The Macmillan Company, 1969.

. . . provides up-to-date and complete information on any writing problem from the smallest detail of punctuation to the most complicated document.

Rathbone, Robert R., *Communicating Technical Information,* Reading, Mass.: Addison-Wesley Publishing Company, 1966.

A guide to frequent uses and abuses in scientific and engineering writing.

Skillin, Marjorie E., Robert M. Gay, et al., *Words into Type,* New York: Appleton-Century-Crofts, Inc., 1974.

Sets forth succinctly, clearly, and fully, the present-day rules and standards of usage covering every step in the preparation of printed material from the manuscript to the finished product.

Strong, Charles William, and Donald Eidson, *A Technical Writer's Handbook,* New York: Holt, Rinehart and Winston, Inc., 1971.

An organized and thorough treatment of style, format, techniques for evaluating data, graphic aids, oral reports, technical advertising and technical editing.

Strunk, William, Jr., and E. B. White, (2nd ed.), *The Elements of Style,* New York: The Macmillan Company, 1972.

A small book, but deservedly well-known for its elementary rules governing usage, composition, form, vocabulary and style.

Style Book for Writers and Editors, New York: The New York Times, 1962.

For everyone who communicates by means of the written word. It enables writers and editors to achieve greater clarity and accuracy through consistency of spelling, capitalization, punctuation and abbreviation.

Tichy, H. J., *Effective Writing for Engineers-Managers-Scientists,* New York: John Wiley and Sons, Inc., 1967.

Helps the professional meet every problem of his working day . . . a most effective and useful desk companion.

Trealease, Sam F., *How to Write Scientific and Technical Papers,* Cambridge: Massachusetts Institute of Technology Press, 1968.

A succinct treatment on preparation of illustrated papers and reports, primarily directed toward theses and dissertations, but applicable to other types of technical papers.

United States Government Printing Office Style Manual, Washington: U.S. Govt. Printing Office, 1973.

Official guide for form and style of government printing.

Ward, Ritchie R., *Practical Technical Writing*. New York: Alfred A. Knopf, 1968.

A practical guide to the principles of technical writing for students of engineering and natural sciences. . . .

Glossary

DEFINITIONS OF TERMS

Most of the terms defined below are used both popularly and technically. Technically used, they are clear and definite; popularly used, they are often obscure and vague. It will be worthwhile to understand their technical meanings.

Facts

A *fact*, according to *The American Heritage Dictionary*, is

> 1. Something known with certainty. 2. Something asserted as certain. 3. Something that has been objectively verified. 4. Something having real, demonstrable existence.—© 1979 by Houghton Mifflin Company. Reprinted by permission from *The American Heritage Dictionary of the English Language*.

Since logically nothing is known for certain, and many things asserted as certain assuredly are not, a fact, for scientists, is something objectively verified or having real, demonstrable existence.

Statements about what you are thinking, what you dream or imagine, or how you feel, may be absorbingly interesting to speakers and their listeners or to writers and their readers. But such statements usually have not been objectively verified. Scientific facts concern the world outside the mind, and have been brought to mind through the eyes, the ears, the nose, the tongue, and the sense of touch. Statements concerning the operation of the mind itself, however, if made by trained psychologists, may be scientific as the data of physicists or plant pathologists.

Obviously the messages brought to the mind by the five senses are not equally significant. Chemists at work over their test tube will be unaware of most of their sensations. They may know vaguely that it is snowing outside, that a bunsen burner is hissing, that a colleague is humming a tune, that a liquid in a beaker smells like rotten eggs. But of none of these facts—facts because they are verifiable, though for scientific purposes not worth verifying—are they really aware. Only the data of the experiment require conscious attention. Scientists learn to ignore sense impressions that do not concern their purposes.

Some examples may serve to illustrate the different kinds of facts.

1. This elm tree is dead.

The statement is a specific verifiable fact. For most people it would be merely an idle observation, but to a student of Dutch elm disease it may be a basic datum.

2. This block is 1.245 inches square.

Like the first example, the statement is a specific verifiable fact. It is based on accurate measurement: not 1¼ inches, but 1.245 inches. Strictly speaking, no such statement is absolutely true, because even with the finest precision instruments there is a limit of accuracy. For most practical purposes the statement is factual.

3. This drop of water is composed of hydrogen and oxygen in the proportion of two atoms of hydrogen to one of oxygen.

For the chemist who has just completed the quantitative analysis of a drop of pure water, Statement 3 would be a specific report of an observed and verifiable fact. If "*a* drop" is substituted for "*this* drop," however, the statement applies not to just one drop, but to all drops. Such inductions may or may not be facts. Chemists now know this one to be not necessarily a fact, since "heavy water," with different chemical compositions, has been discovered.

4. Herodotus lived in the fifth century B.C.

Here is a *historical fact,* based on the observation of persons long since dead. A large part of the knowledge of the past is derived from written records, sometimes made by an observer, sometimes at second-, third-, or fourth-hand. Of the history of the past—of everything outside our own experience—we must be content with such records. We regard the statement about Herodotus as a fact because competent scholars agree that the evidence supporting it is adequate. Most of us accept such statements as factual just as we accept the statement that Hiroshima was destroyed by an atomic bomb. We weren't there to see it, and perhaps haven't even talked with anyone who was there; but the evidence is adequate.

The mere printing of a statement in a book, however, even if in the *Encyclopaedia Britannica,* does not make it a fact. Secondary sources are often unreliable or incomplete or misleading, and even the most careful of historians can make mistakes. Scientifically minded persons will make sure either that they have seen for themselves or that their sources are reliable.

Abstract and Concrete Terms

A *concrete* term designates something that exists in the physical world. It may be *particular* (specific)—that is, designate only one object, such as Blair House, Pike's Peak, Missouri River. Or it may designate a *class* of objects including many particulars: *house, mountain, river.*

An *abstract* term designates something that does not exist in the physical world: *justice, beauty, truth.* Abstractions are concepts derived from a logical process in which the mind "draws from" members of a class, qualities common to all but not existing concretely in all. Thus the statement "The Missouri River empties into the Mississippi" is true, though its truth is merely a quality of the statement. The word *truth* has no physical referent. A specific incident may be an example of justice, but justice is a concept existing only in the mind.

Even concrete class names are "low-level" abstractions. There exists no such thing as "cow"—only cow_1, cow_2, cow_3, etc. Semanticists call such terms low-level abstractions because they include *many* features common to all members of the class. "Animal" is a high-level abstraction because it includes fewer features common to all members of the class. All cows, but not all animals, have four legs and four stomachs.

The ability to abstract and to think in abstract terms is an incalculable advantage to men. Without it, our science would be restricted to particular concrete statements; with it, theories and general laws can be conceived. But failure to distinguish between abstract and concrete terms often results in confusion and endless arguments. Disputes about the reliability of "the press" (an abstraction) are a waste of time; an argument about the truth of a particular statement in the *Chicago Tribune* may perhaps be settled.

General Statements and Laws

A general statement notes a common characteristic of several particulars. Though the inductive procedure of science begins with particular facts, it ends with generalizations. The recording of specific facts would have little value if no general statements resulted.

General statements may be *limited* or *unlimited.* The first is based upon observation of *all* the particulars.

1. All the sheets of paper in my tablet are yellow.

If the meanings of the words in the statement are agreed upon, the general statement is certain and complete. It is limited because it includes only a few particulars.

2. Stones fall toward the earth when dropped.

Statement 2 is an unlimited general statement, based on observation of only a very few of all the particulars included. There are uncounted millions of stones that no one has ever seen fall to the ground when dropped. Yet we accept the generalization without question because competent observers have never reported an exception. We infer that what is true without exception of an infinity of particulars is true of all of them.

On this kind of evidence most scientific knowledge is based. The "laws" of science are such unlimited generalizations. The law of causation—every change in nature is produced by some natural cause—is generally accepted because reputable observers have reported no exceptions. Believers in miracles might question the validity of the law.

Limited general statements, if made by competent observers, are facts. Unlimited general statements, including laws, since they go beyond experience, may or may not be facts. Further observation of particulars may prove them to be false or only partially true.

Hypotheses and Theories

Types of general statements not as widely tested as laws may be referred to as *hypotheses* or *theories*. A hypothesis is a tentative generalization made after observation of only a few particulars. Further observation may prove it to be untenable. It seems to explain certain characteristics that the particulars have in common, and serves as a guide in the observation of others. The history of speculation about the nature of matter, from the first hypothesis by the Greek Democritus, through its modifications by Dalton, Avogadro, Thompson, Rutherford, Bohr, Schroedinger, and others, furnishes an interesting example of the way hypotheses aid in research and then are discarded or modified as studies continue. Such generalizations are often called *working hypotheses.*

When a hypothesis has been thoroughly tested and is generally accepted, it may become a *theory;* that is, a wealth of evidence supports it but not enough to give it the force of law. The same generalization may be variously referred to as "a mere hypothesis," "just a theory," or "an established law," depending upon people's opinions of the weight of the evidence supporting it.

Darwin's generalization about the origin of species is a good example. Originally a hypothesis, it is now generally, but not universally, accepted as the most plausible explanation of the diversity of biological forms, and therefore is commonly referred to as the *theory* of evolution. Some people prefer the opposing notion—there is no scientific evidence to support it, and therefore it cannot rightly be called a theory—that the various species of animals and plants were separately created.

Another kind of generalization, of little importance in science except as a preliminary to forming a working hypothesis, is the *guess.* It is usually based on fancy and imagination and not on careful observation. We may say that next winter will be a warm one, or that catching a cold is the result of sitting in a draft, or that the people of the United States want more interstate highways; but unless we are meterologists or doctors or have systematically sampled public opinion, our generalizations are nothing more than guesses.

Of course the guesses of specialists concerning something in their fields will have more weight than someone else's. The guess of a civil engineer that a proposed dam will be practicable and profitable may have a broad background of experience to support it. Unfortunately, however, persons entitled to make guesses in their own field and be listened to with respect make guesses too often about matters of which they know little. When a noted chemist discourses on economics, or a noted economist on religion, they are likely to make guesses that get much publicity but have no foundation. A necessary part of the training of scientists is to learn to exclude guesses—except as working hypotheses—from their work.

Opinions and Judgments

Opinions and *judgments* (often called *value judgments*) are particular or general statements of varying degrees of validity. Often they are merely uninformed guesses: "We're sure to have a depression soon"; "The governor is dishonest." At their best they are based on wide experience and extensive knowledge. We may have the expert opinions of a lawyer or a doctor or an engineer; the judgments of a court or a board or a jury. In these special senses the opinion is merely offered and may not be acted upon; the judgment usually results in action of some kind. In popular usage, however, the terms are almost synonymous.

The important thing to remember is that neither opinions nor judgments are facts, and that they must be clearly distinguished from facts in scientific writing. To say that Smith took Brown's umbrella from the cloakroom is to state a fact if the evidence is adequate; to say that Smith is a thief is to make a value judgment: Smith has stolen and will steal again. To say that butter costs so many cents a pound is to state a fact; to say that butter is more palatable than oleomargarine is to make a value judgment that cannot be supported by scientific evidence. Even if 99 persons out of 100, or 100 out of 100, agree with the statement, it is still a judgment and not a fact.

A fact is verifiable; there is objective evidence to support it. A judgment, though it may be supported by objective evidence, is not verifiable in the same objective way. Often it is based on emotion, prejudice, or personal taste.

Inferences and Implications

Loosely used, *inferences* and *implications* are almost synonymous. As terms of logic, however, they name different procedures. An inference is a conclusion derived from observation of particular facts. The correctness of the inference depends on the completeness of the observation. We may observe that the street is wet and infer that it rained last night—and be wrong because our inference is based on only one observation. From the same fact we could infer that a sprinkling wagon has passed.

In the inductive methods of science, inferences should be based on adequate evidence and should not be confused with facts. In an experiment with fertilizers it may be a fact that Plot A was fertilized and produced 100 bushels of corn, and that Plot B was not fertilized and produced 70 bushels of corn. That the greater production of Plot A was due to the application of fertilizer is an inference, however, the validity of which depends on the extent to which all the other factors that might have affected production have been kept uniform.

An implication, however, is a conclusion based not on particular facts, but on indirect evidence. Something is implied when it is hinted at by the writer, but not explicitly stated. It is a meaning not directly expressed in words. Sometimes the implication is intentional; sometimes not. From an implication, a reader may make an inference. Mrs. Doe may *say* that she believes you, but *imply* that she does not. From the implication you may *infer* that Mrs. Doe is no friend of yours.

In scientific writing, intended meanings should be clearly expressed, not implied.

Induction and Deduction

Induction is the process of deriving general statements from particular observations. The classical example "All men are mortal"—that is, all men die—is such a general statement derived from innumerable particular observations: Man_1 died, man_2 died, man_3 died, etc.

Deduction is the process of appyling to a particular case a general statement known or assumed to be true. "All men are mortal; Socrates is a man; therefore Socrates is mortal" is a typical example of the deductive *syllogism*, with its major premise, minor premise, and conclusion.

In its simplest from an induction is based on an examination of all the particulars included, and is therefore a fact. "All members of this club have paid their dues" is an example of such a limited generalization. In science, however, the most useful inductions are unlimited generalizations, based on the observation of actually only a small fraction of the particulars included. "All men are mortal" is typical in that it includes millions of individuals not yet born. "All living cells contain protoplasm" is an induction based on actual observation of only an infinitely small fraction of all living cells.

To call such an induction a "natural law" is really misleading. A law is prescriptive, something that must be obeyed; but in science it is merely descriptive. A single exception to the induction will invalidate the conclusion; and in the enormous gap between the number of particulars actually observed and the almost infinite number included in the induction, many exceptions may turn up.

Logically, unlimited inductions are indefensible. They go far beyond the evidence. Were it not for our faith in the order of nature they would be impossible. We *assume* that nature acts according to law; we work not to establish that nature *has* laws, but to find out what those laws *are*. Of course we have evidence for the assumption. But it is well to remember that our faculties and tools for observation are limited, and that statements of nature's laws may never represent ultimate truth.

In deduction we begin with a general statement—either an assumption or an induction. Since assumptions are statements without proof—like the axioms of geometry—and unlimited inductions go far beyond the evidence, it follows that deductive reasoning also is logically open to question.

The very possibility of scientific knowledge, then, depends on two basic assumptions, neither of which can be proved: first, that there is order and consistency in nature; second, that from the examination of a few particulars we can know what is true of all members of a class. The scientist must have faith that both these assumptions are valid.

Fantasies

Unlike facts, *fantasies* come from the imagination and not from the senses. Actually everything in one's mind came originally from the senses, but fantasies are not reports of such experience. The reader usually knows that they are from the imagination, from a make-believe world.

Even children are aware that Alice's "Wonderland" is different from their everyday world, but they willingly accept the fiction for the fun of it. When Milton wrote

> Sonorous metal blowing martial sounds:
> At which the universal host up sent
> A shout that tore Hell's concave, and beyond
> Frighted the reign of Chaos and Old Night

readers know that the poet is describing not a sound actually heard, but one imagined, and willingly permit their fancy to follow Milton's.

Usually the difference between *fact* and *fantasy* is immediately evident to discerning readers. Misinterpretation may result if they are confused. Jonathan Swift, for example, published "An argument to prove that the abolishing of Christianity in England may, as things now stand, be attended with some inconveniences, and perhaps not produce those many good effects proposed thereby." If his readers had not known that Swift was a clergyman who believed the Anglican Church as important as the Crown itself, his satirical intent might have been missed. Satirists may depend on the reader's proneness to confuse fantasy and fact.

Index